Builders and
Building Workers

Builders and Building Workers

P. W. Kingsford

EDWARD ARNOLD

First published 1973
by Edward Arnold (Publishers) Ltd
25 Hill Street
London W1X 8LL

ISBN 0 7131 3302 3 Boards Edition
ISBN 0 7131 3305 8 Paper Edition

Printed in Great Britain by
Billing & Sons Limited, Guildford and London

Contents

Preface

This book is about the men of all kinds in different countries, who have made building and buildings what they are today. It tells the story of the builders in Britain, United States of America, France, Germany and Italy who have contributed to the present scene.

Long before Britain became an industrial society building was an ancient craft directed by master craftsmen. This was all changed by the time of the industrial revolution. Building was directed by master builders who were more businessmen than craftsmen; architects had become professional men: the craftsmen had begun to protect their interests by forming clubs, societies and trade unions. Building became a part of big business which many builders tried unsuccessfully to enter. At the same time the craftsmen and labourers developed into a strong section of the labour movement. The result of all these activities was that Britain was covered with towns and cities.

Their appearance, or look, changed frequently according to what each age thought was pleasant to see, or could afford. By contrast the methods, materials and techniques of building remained almost the same until a hundred years ago when basically new ones began to come into use. Both style and technology were always international. In this book the men whose work is described include eight Englishmen, two Americans, three Frenchmen and a German.

Building has aimed at improving human life and standards of living. But as each individual tried to meet his own needs and men searched for quick profits, it defeated its own purposes. Health and the environment suffered. Slowly society came to realise this. And slowly, too, society worked out a way of planning building in the interests of its members so that building could achieve its real purpose.

1972   P. W. K.

Acknowledgments

I am much indebted for information to the following:

Constructional Engineering Union; Dove Brothers Limited; Hertfordshire County Architect; Holland, Hannen & Cubitts; L. G. Mouchel & Partners; Taylor Woodrow Service Limited; Mr. Edward Trollope; Union of Construction, Allied Trades & Technicians; and colleagues at the Hertfordshire College of Building, as well as the librarian there, who helped me more than they knew. I must also thank the librarians of The Cement & Concrete Association; The Hatfield Polytechnic; the Royal Institute of British Architects, and the St. Albans College of Art.

I am also grateful for permission to include extracts from the following sources:

C.T. Branford Co.: Bayer & Gropius, *Bauhaus 1949–1928*.
Dover Publications Inc.: L. H. Sullivan, *Autobiography of an Idea*.
Hodder & Stoughton Ltd.: V. H. Markham, *Paxton and the Bachelor Duke*.

Other works from which minor extracts have been quoted are as follows:

Allen & Unwin Ltd.: J. Summerson, *John Nash*; E. R. Pike, *Human Documents of the Age of the Forsytes; Human Documents of the Industrial Revolution* and *Victorian Golden Age*; J. Simmons, *St. Pancras Station*.
Amalgamated Society of Woodworkers: T. J. Connelly, *The Woodworkers*.
Amalgamated Union of Building Trade Workers: W. S. Hilton, *Foes to Tyranny*.
American Concrete Institute Proc.: Vol. 25, 1929.
Architectural Press Ltd.: S. Giedion, *Walter Gropius*; G. F. Chadwick, *The Works of Sir Joseph Paxton*.
Architectural Record, August 1969.
Builder, 4th March 1876 and 5th April 1884.
Building News, 1874.
Chambre Syndicale Nationale des Constructeurs en Ciment Armé: A. Lahure (Trans). *Country Life*; T. Davis 'John Nash'.

J. M. Dent and Sons Ltd.: C. B. Purdom, *The Building of Satellite Towns*.
Dover Publications Inc.: Mumford, *The Brown Decades*.
Evans Bros. (Books) Ltd.: N. Davey, *Building in Britain*.
Faber & Faber Ltd.: E. Howard, *Garden Cities of To-morrow*.
Ferro-Concrete: W. Noble Twelvetrees, 'François Hennebique'.
H.M.S.O.: S. B. Hamilton, *Note on the History of Reinforced Concrete*.
Illustrated London News, 1884.
Institute of Civil Engineers, E. Freyssinet, *Pre-Stressed Concrete*, 1949.
Leicester University Press: H. J. Dyos, *Victorian Suburbs*.
Longman Group Ltd.: Hammond, *Town Labourer*.
Manchester University Press: D. MacFadyen, *Sir Ebenezer Howard*.
National Federation of Building Trade Operatives: R. Postgate, *Builders History*.
Oxford University Press: S. Giedion, *Space, Time and Architecture*.
Pelican: Hammond, *The Bleak Age*.
Penguin Books Ltd.: Asa Briggs, *Victorian Cities*.
Phoenix House Publications: N. Davey, *History of Building Materials*.
Routledge, Kegan & Paul Ltd.: Ashworth, *Genesis of Modern British Town Planning*.

I am also indebted to the following for permission to reproduce illustrations:

Faber & Faber Ltd. for drawings of plans for a garden city from Ebenezer Howard, *Garden Cities of To-morrow*, ed. F. J. Osborne.
Union of Construction, Allied Trades & Technicians for photograph of A.S.W. banner.
Department of the Environment and H.M.S.O. for view of Osborne House.
B. T. Batsford Ltd. for view of the Quadrant, Regent Street, from D. Yarwood, *Architecture in Britain*.
Victoria & Albert Museum for Paxton's original blotting paper sketch of Crystal Palace.
R.I. Severs Ltd., for view of Impington College.

I Builders and the Industrial Revolution

The industrial revolution had a great effect on building, and in the early nineteenth century there was a sudden increase in new buildings of all kinds. The reasons for this are clear. The rich and the well off people became richer, while the middle class of professional people and business men expanded greatly. All these wanted fine new homes and they could pay for them. As for the working class who streamed into the new factory towns their needs were met by the jerry builders.

The aristocracy, the great lords with £10,000 a year, and the gentry, the baronets, knights and esquires with between £1,500 and £4,000 a year increased their incomes from land. Their rents swelled because of the growing demand for farm produce for the towns and for which in turn their land was used for building. Bankers, merchants, contractors with incomes between £1,000 and £3,000 a year gathered more wealth from the empire and the long wars with France. In a slightly lower income bracket were the new rich, the leading manufacturers, like Robert Peel who employed 15,000 cotton workers, shipbuilders, warehousemen, and who became rich from the rapid growth of foreign trade. In addition the wars brought about a large leisured class, the investors in government loans. The national debt rose sharply and with it a quarter of a million fund holders. All these were the people growing in numbers and wealth who demanded new town houses. They developed the pleasant habit of seaside holidays and wanted smart and comfortable homes by the sea. Even then stockbrokers began to commute from Brighton.

Below the wealthy were the farmers, shopkeepers, skilled artisans of all trades—building, engineering, clothing, shoemaking—and below the poverty line at an income of about £55 a year were the poor, farm workers, domestic servants, miners, vagrants and the million or so paupers, who received relief from the parish.

Building followed these social changes. The clients and customers were there in large numbers. Architects now had a proper professional training. Capital flowed into the building industry and the general building contractor became important. To meet the demands of the wealthy, contractors and architects planned estates on a large scale, as well as smaller developments of squares, crescents and gardens. All these sprang up rapidly in London, Cheltenham, Tunbridge, Wells, and the seaside resorts, Brighton, Hove and Hastings. The outstanding men responsible were John Nash, and Thomas and William Cubitt.

JOHN NASH (1752–1832)

A great deal of Nash's work can still be seen near Regent's Park in London. His birthplace is not certain but it was probably Neath, Glamorganshire. The third and youngest son of a Welsh mother and English father, Nash spent his childhood in Lambeth on the South Bank of the Thames where his father, William, was a mill-wright. The mill-wright was an aristocrat of labour, well paid and with a strong trade union. 'An engineer and mechanic of high reputation, he could handle the axe, the hammer and the plane with equal skill and precision; he could turn, bore or forge with the despatch of one brought up to these trades. Generally he was a fair mathematician, knew something of geometry, levelling and mensuration. . . . He could calculate the velocities, strength and power of machines, and could draw in plan and section, and could construct buildings, conduits or water courses in all forms and under all conditions required in his professional practice,' William Nash probably made a comfortable living but his brothers Thomas and John (uncles of John Nash) were more successful. John was an apothecary and doctor whereas Thomas flourished as a calico-printer and merchant in Lambeth, with his own mills and fine house.

When Nash was eight his father died, but as soon as he was old enough he entered the office of an eminent architect, Robert Taylor, as an articled pupil with the help of his uncle Thomas. He worked there for eleven years. During those years he had a thorough training in the classical Roman style of building, started by Inigo Jones, from his master who designed many important buildings in London. From Lambeth he could see across the Thames the new Somerset House being built in that same style on the opposite bank.

Nash's first big chance came when he was twenty-six. His wealthy uncle died childless and left him £1,000. He seized the opportunity to leave Taylor's office and start up on his own. But it was not as an architect although he had Taylor's reputation behind him, but as a speculative builder. As a builder usually had a trade, he called himself a carpenter.

Nothing is known about his business activities until after a year or two he bought some property in Bloomsbury. This property is at present the premises of The Pharmaceutical Society at the corner of Great Russell Street and Bloomsbury Square. When Nash bought it it was a plain brick building of the early eighteenth century consisting of three houses, 17 Bloomsbury Square and 23 and 24 Great Russell Street. He then altered its appearance so as to make it more imposing. He covered the outside with stucco and on the Bloomsbury Square front added eight large Corinthian pilasters and a heavy cornice and rusticated the ground floor to make it look like stone. The use of stucco on such a scale was itself a new thing. In this now impressive property Nash occupied one house and let the others.

However, he must have been too ambitious for soon after, in October 1783, he was declared bankrupt. At the age of thirty-one his first venture as a London builder was ended and he had to make a fresh start somewhere else. He turned to Wales, his mother's native land, and settled in Carmarthen where he may have joined her.

Carmarthen was a county town of some importance and prosperity being a port on the river, a stage post on the coast road and a market for the whole district. There were prospects for an energetic builder. Nash quickly began business; in 1784 he went into partnership with a carpenter named Saxon and the following year they obtained a large contract. This was for a new roof and ceiling for the old church of St. Peter's, Carmarthen. Their tender was 600 guineas. Nash and Saxon as carpenters bargained with the other tradesmen, such as masons, whose work was necessary. They built several houses in the town, including one for Nash himself in Orchard Street, a plain simple building. On the side he took a lease of some land with a lime kiln, and bought a large piece of woodland as a speculation in timber. Nash who was a very sociable man, made some influential friends among the local squires. One of them, John Vaughan, M.P., commissioned him to design and add a bathroom to his house. Nash said that this was the 'first money he received as a professional man'.

However, his first real start as an architect came from something much bigger. This was Carmarthen Gaol 1789–1792. A few years earlier John Howard, the reformer, had begun his campaign against conditions in the prisons and published his *State of the Prisons in England and Wales*. Carmarthen accepted his criticism and decided to build a new gaol. Nash designed it, using the lessons he had learnt in Taylor's office. Straight away he was commissioned to design a second prison at Cardigan which had also been severely criticised on account of absence of water supply, the filth and vermin. The new prison was described by a contemporary as:

'A very handsome Building, and on a good Plan . . . it has a Chapel which is kept remarkably clean: the Central or Inspection Room, is the Residence of the Gaoler. One great disadvantage, however, is the want of a ready supply of Water, none being to be had but what is fetched by the Keepers from a Well at a distance of a quarter of a Mile.'

Having acquired a name for prisons, Nash went on to design a third one, at Hereford which was finished in 1796 at a cost of £18,647. He had a commission of 4½% on this and travelling expenses of £340. All these gaols were successfully designed on the classical lines followed since the days of Inigo Jones. In the meantime, however, Nash was asked to rebuild the west front of St. David's Cathedral, the wall of which was being pushed out by pressure from the older massive arches inside. This was in the medieval Gothic style and as Nash said later, 'I hate this Gothic style; one window costs more trouble in de-signing than two houses ought to do'. His own description, also later on, of what he did was:

'I constructed a timber buttress as a temporary foundation, and sup-ported the whole of the Saxon arches; I then took down the wall from top to bottom, prepared a new foundation and rebuilt the wall upright, encased the temporary buttress with stone, and substituted stone arches for the wooden buttress. . . .'

In fact he did much less than this and it all had to be done again about sixty years later by Sir George Gilbert Scott (see Chapter 3).

Through the influence of John Vaughan he next obtained a commis-sion at Abergavenny. That little Welsh town wanted, in the spirit of the age, to improve its appearance. In 1793 the vestrymen decided 'that Mr. Nash, architect, should be sent for to take a view of the old Market House, to hear the different opinions of the sixteen Persons appointed, two from each ward, in regard to the different situations and estimates and expenses for erecting the market house upon, to value the old Materials and to point out what will be the probable expense for Paving, Lighting and Draining the Town'. There were also to be 'Butchers' shops, shops, sheds, stalls, and shambles and other conveniences as shall be judged proper according to a plan and estimate to be delivered in by Mr. Nash'. The carpenters' estimate of £810, recommended by Nash, was accepted, and Nash himself received his fee of £52.10s.

Meanwhile, Nash continued to build up a reputation. His circle of friends and acquaintances among the landowners and notables widened. He became prominent in local society. Always an actor, he took a leading part in amateur theatricals in Carmarthen. In a production of 'The School for Scandal' he played the part of Sir Peter Teazle, while the other parts were performed by local leaders of fashion. These contacts led to commissions to build a number of houses for the land-owning gentry, in Pembrokeshire near Boncath, at Haverfordwest; at Tenby and near Maenclogoch; in Monmouthshire, near Newport; in Cardiganshire, near Aberayron; at Cardigan and Henllan Bridge. All these houses, eight or so, were designed and built in two or three years between 1792 and 1794. All of them, too, were in the English classical style which Inigo Jones had bought from Italy, 'combining the advantages of an English arrangement with the beauty of a Palla-dian plan'.

Having made a name in Wales, Nash decided to return to London. Accordingly in 1796, now aged forty-three, he moved into Duke Street in the fashionable district of St. James's. He now quickly gained a national reputation as a designer of country houses for the gentry and the wealthy middle classes. His clients included nobility, lawyers, bankers, politicians and country gentlemen. Many of them were mak-ing money from the rapid growth of industry and trade, and some from the war. Three years earlier the twenty-year-long war with revolutionary France and with Napolean had started.

At first Nash owed his success to his partnership with Humphrey Repton, an expert in landscape-gardening. Thanks to the prosperity of the landowning class Repton came to the fore as a consultant in improving and making more picturesque country estates. His schemes frequently included new mansions, and these he handed over to Nash who designed them in the fashionable imitation castle style. Nash paid Repton 2½% out of the 7% he charged Repton's clients; for his own clients the charge was 4½ or 5%. In the next ten years he built many great houses in different counties, half a dozen in Ireland, and even in Scotland about which he wrote to a friend; 'Just returned from ye cursedest journey to Scotland I have ever experienced'. According to his account he had 'travelled in the three Kingdoms Eleven thousand miles in the year and in that time he had expanded £1,500 in chaise hire'.

Two years after returning to London Nash took another step to fame. He came to the notice of the Prince of Wales, the future Prince Regent, by designing him a conservatory. But more important, he married Mary Ann Bradley, the beautiful daughter of a coal merchant; he was forty-six, she twenty-five. Mrs. Nash was said to be one of the mistresses of the Prince, either before her marriage, or after, or both. Twenty years later, the following ballad was published at a time when the Prince, then George IV, wished to divorce Queen Caroline; 'she' in the poem is Caroline:

> 'She's the people on her side, and that nobody can doubt,
> Like a torrent in her cause they incessantly flow,
> The dirty tricks of G . . . ge . . . oh! at last they are found out,
> What the Devil I'm to do—only God himself can know.
> > In the boat with Mrs. N–sh,
> > To the Isle of Wight I'll dash,
> In my cabin smoke and drink, and forget I have a foe;
> > There's C. . . . . . e my Wife,
> > She's the plague of my life,
> And from her I'll make sail with a yo heave ho.

> Now the Yacht it is all ready, for the Isle of Wight I'll steer,
> With the lovely Mrs. N–sh I am all upon the go,
> I'll pack the husband off—for he has no business here,
> With his Wife in pleasure's bark, I will jovially row,
> > In the Place of Waterloo,
> > He may fume, fret and stew,
> 'Tis an architects disgrace, and that the world does know,
> > Here at Cowes I will revel,
> > Kicking Virtue to the Devil,
> With N–sh have a Smash, singing yo heave ho.'

Whatever the truth of the matter Nash from then on had far more money than his practice would draw in so that people wondered where it came from. He was able to buy not only a grand mansion in exclusive Dover Street, London, but also an estate in the Isle of Wight, where he built an enormous residence with towers and turrets, East Cowes Castle.

In 1806, the year after Trafalgar, Nash took yet another step towards his greatest achievement. Through the influence of politicians connected with the Prince of Wales he was appointed architect to the Chief Commissioner of Woods and Forests, a Government department. For such a prominent man it was only a minor post at £200 a year but it made it possible for him, four years later, to seize his greatest opportunity—the planning of Regent's Park and the grandiose way right down to Carlton House Terrace at St. James's Park.

This opportunity for town planning on a grand scale arose in this way. The Crown estate known as Marylebone Park was about to revert to its owner as leases ran out. It was an area of meadowland with three farms, some cottages and two public houses, the Jew's Harp and the Queen's Head, stretching from Marylebone Road (then called the New Road) north to Primrose Hill. It was a piece of land 'over which', a government official wrote, 'it is probable that on the return of Peace, the town may be expanded'. It was clear that the estate was ripe for development, and therefore in October 1810 the commissioners of Woods and Forests told its architect (Nash) and the architects of Land Revenue the other government department concerned, each to prepare designs and reports. There was little doubt that Nash would win. In February 1811, the Prince of Wales became Prince Regent because of his father's madness. In March, Nash was talking freely about his plans for the estate. By July, the plans were submitted and in October the Treasury approved Nash's scheme; the Prince was 'so pleased with this, magnificent plan, that he has been heard to say "It will quite eclipse Napoleon" '.

Nash's plan was a thorough one, including sewers as well as building roads and planting trees, canal, a park and a great thoroughfare all named after the Regent. The park was to be surrounded by terraces of grand houses and designed so 'that the attraction of open space, free air, and the scenery of Nature, with the means and invitation of exercise on horseback, on foot, and in Carriages, shall be preserved or created in Mary-le-bone Park, as allurements and motives for the wealthy part of the Public to establish themselves there'. This area was to be linked southward with Portland Place which already existed.

As he pointed out 'the best built part of that quarter of the town is comprised between Baker Street, Westward and Portland Street, Eastward; Mary-le-bone Park very fortunately lies immediately behind those streets and the Crown has therefore the power of preserving that best built part of the town from the annoyance and disgrace which threaten it on either side, and of establishing a beautiful termination to that elevated and conspicuous Boundary of the Metropolis'. Here he built Park Square and Park Crescent.

The line southward to St. James's Park could not be continued straight from Portland Place and so Nash positioned All Soul's Church, Langham Place, so as to ease it into Upper Regent Street and then by means of a circus cross Oxford Street. Regent Street itself Nash planned as partly residential, partly commercial, sweeping round in The Quadrant at its southern end to cross Piccadilly by a circus. This new street was authorised by a Bill in Parliament in 1813 called, 'An Act for making a more convenient Communication from Mary-le-bone Park and the Northern Parts of the Metropolis . . . to Charing Cross within the Liberty of Westminster; and for making a more convenient Sewage for the same'.

The difficulty of planning a new road in the heart of London was partly social. Nash saw the new Regent Street as a social barrier; he described it as:

'A boundary and complete separation between the Streets and Squares occupied by the Nobility and Gentry, and the narrow Streets and meaner Houses occupied by mechanics and the trading part of the community. . . . It was to cross the eastern entrance to all Streets occupied by the higher classes and to leave out to the east all the bad streets and as a sailor would express himself, to hug all the avenues that went to good streets.'

In this way with a continuous colonade from Oxford Circus to Pall Mall (which was never built) 'those who have daily intercourse with the Public Establishments in Westminster, may go two-thirds of the way on foot under cover, and those who have nothing to do but walk about and amuse themselves may do so every day in the week, instead of being frequently confined many days together to their Houses by rain; and such a covered Colonnade would be of peculier convenience to those who require daily exercise. The Balustrades over the Colonnades will form Balconies to the Lodging-rooms over the shops, from which the Occupiers of the Lodgings can see and converse with those passing in the Carriages underneath, and which will add to the gaiety of the scene, and induce single men, and others, who only visit Town occasionally to give a preference to such Lodgings'.

After crossing Piccadilly the new royal route was to continue in Lower Regent Street with its two theatres, and so to Pall Mall. The creation of Waterloo Place, the building of Carlton House Terrace and the laying out of St. James's Park were planned after Regent Street had proved a success.

View of the Quadrant, Regent Street

Osborne House from the air

The whole plan had to be a success, financially as well as artistically. Nash's reputation and the money poured in by the Treasury to the tune of a million pounds depended on it. Nash himself put many thousands of his own money into it. The whole development, starting in 1812, the year when the Russians burnt Moscow rather than give it up to Napoleon, took nearly twenty years to complete. It was a complicated as well as a large one. Starting from the northern end, Nash designed the terraces surrounding Regent's Park and Park Crescent leading south into Portland Place. This area was comparatively easy to plan as the land had not been built on. But Nash had to negotiate with reliable builders who would not go bankrupt, as one of them did. In addition, there was the new Regent's Canal, along the northern boundary of the new Park, to be constructed as a branch of the Grand Junction Canal. This he launched himself, organising a company to finance it, and subscribing £15,000.

Regent Street itself was a much more difficult task. Seven hundred houses had to be demolished and the owners compensated. Nash dealt with the valuations himself. At the same time he was engineer in charge of the new sewer required. It came under criticism but the contractors declared that it was 'as complete a Sewer, the whole length, as can be executed by the hands of man'. Nash also dealt with the leasing of sites to builders such as James Burton, as soon as the street was open. Moreover, as he did not design all the façades in Regent Street he had to negotiate with half a dozen other architects in order to keep the overall plan.

He was responsible to the Commissioners for the whole finances of the great development. The building of Regent Street began at the southern end. Lower Regent Street, together with Suffolk Street and Place, was built first, between 1817 and 1820. But when Nash came to the Quadrant (1819–1820) just north of Piccadilly Circus he had to plunge in financially himself. The Quadrant was the most admired part of Regent Street. It was one continuous block sweeping round in a full quarter-circle, of over fifty houses with shops on the ground floor and dwellings on three storeys above. Along the front on either side of the Street were continuous curving colonnades, reaching up to include a mezzanine floor, with 145 columns. In order to get this constructed as one unit Nash had to lease the whole plot, supply the money and build it himself. He said later, 'I do not think the quadrant would ever have been carried into execution but in that way'. His method was to give the work to a group of craftsmen on condition that each took a lease of at least one house. '. . . Prigg, the plumber, took two houses; Want and Richardson, bricklayers, seven houses; Brine, who purveyed mantelpieces, one house; Palmer, the glazier, two houses, and so on.' But he had to lend them the money to do this amounting to £60,000. This they repaid by their work.

At last the whole development was completed. Starting in 1810 when Nash made his first designs for Regent's Park, at the age of fifty-eight, it finished with the layout and planting of St. James's Park in 1827, when he was seventy-five. After the long years of upheaval, dirt and disorder in central London foreign visitors as well as Britons admired the result. One of them wrote in 1826:

'London is, however, greatly improved in the direction of Regent Street, Portland Place and the Regent's Park. Now, for the first time it has the air of a seat of Government and not an immeasurable metropolis of "shopkeepers" to use Napoleon's expression. Although poor Mr. Nash (an architect who has great influence over the King, and the chief originator of these improvements) has fared so ill at the hands of the connoisseurs, and it cannot be denied that his buildings are a jumble of every sort of style . . . yet the country, is, in my opinion, much indebted to him for conceiving and executing such gigantic designs for improvement of the metropolis.'

The 'connoisseurs' objected to the way Nash always used stucco. He had in fact used at first Roman cement discovered by James Parker in 1796, and then from 1820, a patent mastic cement. This was linseed oil boiled with litharge and mixed with finely powdered porcelain clay. It had to be applied to the walls as soon as possible after being mixed and then trowelled continuously until it was set. The vast areas of stucco must have meant a tremendous amount of work for the plasterers. A contemporary comment ran:

'Augustus at Rome was for building renowned,
And of marble he left what of brick he had found;
But is not our Nash, too, a very great master?—
He finds us all brick and leaves us all plaster.'

He was by now a wealthy man. He told a visitor at East Cowes castle that as well as the castle and land, which was worth £30,000, he had a farm on the Isle of Wight worth £30,000, 1,100 shares in Regent's Canal worth £100 each, and an income of more than £7,000 a year.

Two other large works of Nash must be mentioned, both of them for his royal patron. The first was the Pavilion at Brighton to which, as both architect and builder, he gave its present appearance. The work, carried out at the same time as Regent Street, took eight years and cost £160,000. The other was Buckingham Palace. Nash was seventy-five when he began the rebuilding of the Palace to the King's command. The extravagant expenditure caused much resentment by the public and he was criticised for his mistakes. When, however, he was charged in Parliament with making dishonest profits from property deals a select committee cleared him of any blame. The King wanted to reward his architect and wrote to the Prime Minister, the Duke of Wellington:

'My Dear Friend,
        I now write to you upon a matter in which I feel very much interested. The Report of the Committee of the House of Commons upon Mr. Nash's business has been delivered in, and, as I am informed by one of that Committee (not one of those who had any previous predilection towards him) "without the slightest stain or imputation upon or against his character," I do therefore desire that you will direct his being gazetted by himself on Tuesday next the 16th of this month, as a Baronet, with the remainder at his death (as he, Nash, has no family of his own), to his nephew Mr. Edwards, a gentleman of excellent character, large property, who sat in the last Parliament, and who has proved himself a thorough supporter of government, and a most loyal man, besides being well known to me personally. Mr. Nash has been most infamously used, and there is but one opinion about it; and therefore it is not only an act of justice to him but to my own dignity, that this should forthwith be done. For if those who go through the furnace for me and for my service, are not protected, the favour of the Sovereign becomes worse than nugatory.
                Your very sincere friend
                                G. R.'

The Duke, however, would not agree. George IV died in June 1830 and in October Nash was dismissed and work on the Palace was stopped.

Nash was now seventy-eight and he retired to his castle home at East Cowes. He had always been a very active, busy man. Here is an account of a conversation with him: 'He spoke of his health which was generally good but is occasionally subject to pain in the right side, which the Doctor pronounced to be a *dumb colic* (grumbling appendix?). His object is to keep his body open; he then is well. The prevailing disposition is to form Alkali, he therefore takes acids. The liquor Punch agrees with him.' However, shortly before he retired he had a stroke from which he did not fully recover. He designed a church for the village without charge. He noted events in his diary:

'March 22 (1832). London—called on Dr. Johnson and was cupped—drove to the village—went to the play at Adelphi—The Reform Bill passed the Commons by a majority of 110.'

But three years later his strength had failed. On May 7th, 1835, he noted: 'very ill'; on the 11th: 'much worse'. He died two days later, and in debt. His money had vanished. Of his great town-planning scheme, his 'garden city for the well-to-do in the centre of the metropolis', much remains today; the terraces and crescents of Regent's Park, Park Crescent, and All Soul's Church, Langham Place; in Regent Street only the line and a few buildings and the Haymarket Theatre; the United Services Club in Pall Mall.

THOMAS CUBITT (1788–1855)

Thomas Cubitt, like Nash, was born into an age of development of large estates in London. Born at Buxton, near Norwich, a year before the French Revolution, he died during the Crimean War. He had two brothers, William and Lewis, about whom more later. His father, who was a small farmer, died when Cubitt was eighteen. In 1806, the year after Trafalgar, Cubit, who had served at least part of his apprenticeship to a carpenter, was taken on as ship's carpenter on one of H.M. frigates. After a long voyage to India he returned to England three years later and with his savings started on his own as a master carpenter in London.

Fortunately for Cubitt he started in business at the time when a number of big estates in London wee being developed. Partly because of the profits to be made in building houses for the growing middle class, partly because of the inconvenience caused by old buildings, partly because of a wish to make London a more pleasant and attractive place, many great landlords set about planning and rebuilding from the mid-18th century. There was the Bedford estate belonging to the Duke of Bedford, the Portland estate belonging to the Duke of Portland, the plain Mr. Portman's estate, the Southampton estate, the noble Lords Berners' and Somers', and last but not least the estate of the Foundling Hospital. Most of this large area lay roughly between Regent's Park and Oxford Street and between Edgware Road in the west and Gray's Inn Road in the east. It was a great opportunity for builders and architects.

Before Cubitt came on the scene, a very successful builder, a Scotsman called James Burton, had already developed much of Bloomsbury, laying out Russell Square, and the neighbouring streets on the Bedford estate, part of Tavistock Square and Bloomsbury Square. He sometimes employed as many as 200 men. Cubitt followed on after him. In fact the first work we know he carried out, about 1815, was re-roofing one of Burton's building which were not always up to standard. It was the Russell Institution in Coram Street which Burton had built only a few years before. This work he did under an architect, John Shaw, but thereafter as he established his own business he rarely worked with architects. Often he designed his houses himself, or employed his younger brother Lewis who had been trained as an architect. In the same year, a young man of twenty-seven, he obtained his first big contract; to build the London Institution in Finsbury Circus. His tender was accepted on the recommendation of the people who had seen the standard of his work at the Russell Institution. It was a large handsome building with a library, lecture theatre and classical façade, erected like many others of that period to give education for the middle class. It was demolished in 1936.

At about this time Cubitt became a speculative builder; that is he took leases of land and built houses for sale or letting, borrowing money in the City to do so. He started in North London. Soon after 1815 he took a lease of part of the Calthorpe estate, east of Gray's Inn Road and built for example, Frederick Street. At Highbury he built some medium-sized villas, with good gardens, which were soon let or sold. Then at Stoke Newington he bought land used for market gardens and grazing cattle and built terraces and detached villas for which he had to construct the approach roads. Nearby he also bought the freehold of a nursery garden and grazing ground, six acres in extent, called Barnsbury-park. This he laid out as Barnsbury Square, built a few houses and villas as examples and let the remaining ninety sites to other speculative builders. Thus he quickly developed a considerable business.

At the same time he broke away from the traditional way of organising operations and started a new system. Until then the usual practice was for a tradesmen (e.g. a carpenter) who obtained a contract, to sub-contract to other tradesmen (e.g. bricklayers, painters) those parts of the job which only they could do. Cubitt found the disadvantages of this. Faced with completion penalties he was at the mercy of the other tradesmen and had no control of the operations. He therefore engaged all the tradesmen he needed on the basis of regular payment of wages; ganges of carpenters, smiths, plumbers, glaziers, painters, bricklayers, with the foremen for each trade, and these became his employees or wage workers. They were based on the great new workshops and yards, complete with horses and carts and materials, which he built on land he bought on the east side of Gray's Inn Road. The new organisation caused a sensation among tradesmen and architects.

However, it created difficulties. A continuous stream of work was necessary to keep it going. There was the problem of giving such a large body of workmen regular work and of keeping them employed once they had got used to the new system: therefore Cubitt was led into continually leasing land and building on it, as a speculation, on a bigger and bigger scale. A contemporary wrote:

'His plan was to take a large tract of unoccupied land, sometimes from several district landowners, and to lay it out on one great plan of squares, streets and roads, etc., as a whole, sparing no expense in the outset in the drainage, forming gardens, planting, laying out wide streets, and using every endeavour to keep up the character of the whole'.

On the business side, he insisted on regular monthly payments from clients, even when they were royalty, as he did later at Osborne House and Buckingham Palace.

In 1820, when Cubitt was thirty-two, he appeared on the scene of the large Bloomsbury estates. However, at first it was as a contractor rather than a speculative builder. Again he took over where James Burton left off. Burton had built the east side of Tavistock Square. A stockbroker acquired the lease of the south side and contracted with Cubitt to build houses as an investment for his daughters. The houses let well at £150 a year. Encouraged by this Cubitt built the other two sides of the square on his own account. Then in 1824 he bought from the Duke of Bedford and Lord Southampton the 99 year lease of a large piece of land near his Gray's Inn Road workshops. On this over the years he built Woburn Place, Gordon Square, Endsleigh Place, Tavistock, Gordon and Endsleigh Streets and part of Euston Square. He was still building in the Bloomsbury district when he died, thirty years later.

There were plenty of opportunities for development in London. Cubitt seized several so that together, the operations on each often overlapping, they gave a continual and growing volume of work for his firm which lasted through the eighteen-thirties, 'forties and 'fifties. After his Bloomsbury purchase, in 1825 he acquired another large piece of vacant land, a bare swampy-expanse of 140 acres which separated Hyde Park Corner from the village of Chelsea. Known as the Five Fields, it was the haunt of snipe, footpads and fog, but it was near to Buckingham Palace. Cubitt bought the leasehold from Lord Grosvenor and Mr. Lowndes because he noticed that the fashionable world began moving west when the new king, George IV announced that Buckingham House was to be made into Buckingham Palace. There were great possibilities. In 1820 Lord Grosvenor obtained a Private Act for draining, raising and levelling the site. Over the next twenty-five years Cubitt created there a new home for the aristocracy, Belgravia.

This magnificent estate was laid out in splendid squares, crescents and places. Hundreds of stucco-fronted houses in the classical style were built by Cubitt. The show-piece, Belgrave Square, designed by an architect, had been the site of a refuse dump; it consisted of a handsome block on each side and a separate mansion at three of the corners and was not finished until 1856. South of it, the long 'parkway' of Eaton Square, went up between the years 1827 and 1853. Lowndes Crescent was laid out between 1836 and 1849. There was no end to the demand from the wealthy; rents of £1,000 a year for a house and stables were not unknown. Cubitt and his brother Lewis designed the whole expanse, apart from Belgrave Square. The result may still be seen as an example of their work.

Cubitt undertook two further major developments. After Belgravia he laid out the whole area south of it as far as the Thames, known as Pimlico, on a less grand scale, and built his own works on Thames side. In about 1825 he went farther afield, south of the Thames, and leased Clapham Park south-west of the Common, from the Lord of the Manor. Clapham Common, four miles south-west of London Bridge, was the home of City merchants and bankers. Cubitt laid out the whole 250 acres with miles of roads, planted thousands of trees, built mansions and villas for City men and sublet to other builders.

The name of Cubitt, already famous in the building world for new methods, successful speculation and high standards of construction, became more widely known in the eighteen-forties. In 1843 he gave expert evidence to the Royal Commission on the State of Large Towns and Populous Districts appointed by Sir Robert Peel because of the widespread insanitary conditions of living and the danger from cholera.

Here are some of the answers he gave to the Commissioners' questions:

Cubitt: I have built very few houses for poor people, but when I do, I try to make them as healthy as possible.

Questioner: Have you ever built them back to back?

C: Never.

Q: Has it been your habit in the construction of buildings of an inferior class to place water closets in them?

C: I scarcely build any house, however small, without having a water closet attached to it and not a common privy.

Q: The object to be desired would be to substitute water closets for privies?

C: That could be done only as the demand for comfort arose.

Q: In some way you would enforce a good supply of water for the poorer classes?

C: Yes. I would wish the supply of water to be much more extended. One of the great difficulties in being cleanly in and about London is the great quantity of black material constantly floating about in the air. . . . The first thing that ought to be done is to prevent the great smoke of chimneys; they ought to be stopped.

Q: You think it is injurious to trade as well as to health?

C: Yes.

Q: You think it would be quite practicable to prevent it?

C: I think our atmosphere need not be half as bad as it is.

Q: You are of the opinion that if means are provided for promoting personal cleanliness and general cleanliness the poor would be induced to avail themselves of any means of thus advancing their comfort?

C: I think that if you could give them the means of advancing a little in that, and induce them to take the first step they would soon rise two or three steps of themselves. Poor men and rich men are all of the same material and their minds are working to nearly the same points. One of the greatest improvements of London would be to keep the people all well employed; and I do not know anything that would do more for the labouring classes than to give the rich people the power of having their houses very nice as it would induce them to spend more money.

Two years later when Queen Victoria bought Osborne House and a thousand acres in the Isle of Wight, Cubitt was the builder of her choice. He collaborated with the Prince Consort. The general design, in the style of an Italian villa, was by the Prince, and Cubitt made the drawings and carried out the whole work which lasted for six years. 'Both him and the Queen have been very civil', Cubitt wrote to his wife. The Duke of Wellington on a visit there, being told of Cubitt's work said, 'Oh I must see him. He's a great man.'

He was often active in attempts to improve the environment in London. Commissioners for the Improvement of the Metropolis had been appointed by the government. Their first problem was how to get a proper embankment constructed along the Thames from Vaux-hall Bridge to Battersea Bridge. Cubitt gave evidence and helped practically. He held the lease of a river frontage on the Middlesex side of more than 3,000 feet and he offered to build a handsome granite-topped embankment wall along the whole of it at his own expense. The length was from Vauxhall Bridge west to Bakers Lane which ran from the west end of Lupus Street down to the river. The estimated cost was £5 per foot. He was under a legal obligation to build it along only 700 feet. It was a shrewd as well as a generous offer; as another expert pointed out: 'The Embankment is needful to Mr. Cubitt. He has a young property which is now building. This Embankment is useful to Mr. Cubitt. His offer is evidence of the great importance he attaches to it.' Cubitt had of course his own works on the river front-age at that point. All the same his offer was also 'to give the public enjoyment of the frontage and use of the road along it'. He reminded the Commissioners of his own reputation:

Question: You have a considerable quantity of ground in this district?
Cubitt: Yes, my ground is principally in the parish of St. George, Hanover Square, and extends inland to Pimlico. The Commission will, I am sure, have observed that I have laid it out in wide streets and give good communications for the public; and that my lines of thoroughfare are very superior in character to those of the adjoining neighbourhood.

When his offer was accepted Cubitt was then asked if he would construct the whole Embankment from Vauxhall to Battersea Bridge at cost price. He replied on 26th March 1844 from his house in Eaton Place:

'I will do it in the most prompt manner possible, and at the lowest cost . . . the interest I feel and the great desire I have to see this work fully carried out will be a sufficient inducement for me to use all possible means to do it in a most economical manner without any kind of profit out of the work. . . .'

At the same time Cubitt pressed for the creation of Battersea Park on the site of Battersea Fields. He was anxious that the Government should buy the Fields and the unoccupied land along the Thames before prices rose and they were covered with bricks and mortar. To this the vicar of Battersea added his plea:

'The vast number of persons visiting Battersea Fields on Sunday are drunken, lawless and licentious. It is occupied now in growing hay. The cultivation has however been injured by the immense swarm of people that spread themselves over the fields on holidays and Sundays.

'My profession has given me a full knowledge of the close and unwholesome places which are inhabited by the artisans and working men of the South London district. I feel confident that many of these would refrain from the spirithouse if the opportunity were afforded them of enjoying themselves away from their uncomfortable homes. By encouraging healthful recreation the Commissioners will promote social and domestic happiness; they will implant feelings which are now deadened by dirt, drink and discomfort.'

Battersea Fields was bought; but when later on the Chancellor of the Exchequer, Benjamin Disraeli, hesitated to spend money on them Cubitt offered to make a park at cost price.

A few years later he played an important part in the great event of the eighteen-fifties, the Great Exhibition of 1851 in Hyde Park. Two years before Prince Albert called Cubitt with others to Buckingham Palace to begin planning. Cubitt was one of those who guaranteed the money required. Afterwards when the Brompton estate, on which was built the Albert Hall, was bought with the proceeds of the Exhibition, he negotiated the purchase free of charge.

By this time Cubitt had become rich. His works at Pimlico was very large and highly organised for the building industry of that time. When it was burnt down in 1834 the loss was estimated at £30,000. It included a joiners' shop 200 feet long, a marble works four floors high, containing cutting machinery, a cutting-house also four floors high in which timber was cut up before being passed to the joiners, plaster model shops, engineers' works two floors high and 100 feet long, and a large store containing valuable prepared work such as ornamental flooring and fittings for Buckingham Palace.

At the time of his death he had a reputation as a good employer. That was twenty years after the strike at his works, described on page 36. At the time of the Pimlico fire he is reported to have said, 'Tell the men they shall be at work again in a week and I will subscribe £600 myself towards replacing tools they have lost'. The number of workmen was sometimes as many as 2,000, plus clerks, foremen and managers. For them he provided a library and a canteen with cheap soup and cocoa 'to prevent dram-drinking', and schoolrooms for their children. He died worth over a million pounds. To his widow he left £20,000 and an annuity of £8,000, and the rest to his large family.

WILLIAM CUBITT (1791–1863)

The two younger brothers of Thomas Cubitt were smaller men in the building industry but each a big man in his own way. They both developed as assistants to their brother.

Like his elder brother, William Cubitt went to sea when he was young. He served four years in the Royal Navy during the Napoleonic war. In 1810, at the age of nineteen, he left the navy and joined his brother who had just started as a master carpenter in London. From that time he helped him to start and organise the new methods of production at the Gray's Inn Road works. Fifteen years later, when he was thirty-four and his brother was busy with his large-scale planning in Bloomsbury and Belgravia, he took over complete control of the works. Thus the two brothers, as partners, divided the work between them. He was responsible for the actual building and construction while his brother Thomas, concentrated on the financial and planning side as the business man undertaking the risks of speculative building.

William was in charge of the works at 37 Gray's Inn Road at the time of the strike in 1834.

Much of the work which William Cubitt carried out was for his brother's schemes; for him he built large parts of Belgravia and Pimlico. But there were many other important works which he did separately. In 1829–1830 he had a contract of £63,000 for Hanwell, Middlesex, lunatic asylum. He worked for the architect, Philip Hardwick, on two large operations; the warehouses at St. Katherine's Dock, London, in 1827–1828, and at the new Euston Station where he built the Great Hall (since pulled down), 60 feet in width and with coffered ceiling, in 1849. He also designed and built the tall Welwyn Viaduct over the river Mimram for the Great Northern Railway. A few years later the firm had grown so much that he took several partners and made it into a private company—W. Cubitt & Co.

He retired when he was sixty and became very active in public work in the City. The firm continued to grow. A few years later a party of architects went on a visit to the Gray's Inn Road works. 'Most Londoners have heard of Cubitt's', according to a report of the visit. It described the works in detail. First:

'The principal joiners' shop, a very large, lofty and well lighted room, in which all the best joinery work is done. Here were to be seen in various stages of progress door and window frames, doors, partitions, office and other fittings, all the best of their kind, and in various woods, much of the work being for large mansions of the nobility which the firm is now engaged in erecting or enlarging. The manager of the carpentry and joinery department having explained and practically exhibited to the visitors the different modes of veneering, the party entered the "Wood Museum" of the establishment, which is a miniature edition of the Museum at Kew Gardens. The little museum at Cubitt's leads out of the principal joiners' shop, and contains specimens of every description of wood, British, foreign, and colonial, used in the building trade, and others besides. Among the many interesting practical points to which Mr. Wyatt drew attention was that a great deal of the "figure" of the wood depends upon the cutting. Wainscot oak cut in one direction may have no figure, but yet, cut in another manner, may be full of it. In bird's-eye maple, again, the figuring will present different appearances according as it is cut from the outside of the tree or towards the heart. Sections of various casement stiles, including one (made for the Duke of Westminster) of peculiar construction and shod with brass, in order to form an air-tight and dust-proof joint, having been explained, the visitors entered the machine joiners' shop, another very large and well-lighted apartment, where were to be seen in operation planing, mortising, tenoning, and various other machines capable of doing almost, if not quite, everything that can be done in the shape of wood-working. A very ingenious machine for doing all descriptions of "Gothic" work, sunk lozenges, &c., was particularly interesting; indeed, the wood working machinery generally attracted considerable attention, not only on account of the almost infinite variety of work which it is capable of turning out, but because of the excellence and beauty of finish, to say nothing of the marvellous rapidity with which the most elaborate pieces of work are produced.

. . . among the specialities pointed out were a peculiar system of tongued-and-grooved flooring which Messrs. Cubitt have largely adopted, and the patent dowelled doors (made by special machinery), framed together with strong oak dowels instead of mortise and tenon joints,—these doors having been used very largely (if not eclusively) in the Peabody buildings. Having visited the saw-mills, the party passed to the wrought-iron girder shop, where girders were seen in process of building up and riveting, and powerful punching and shearing machines were cutting and holing stout iron plates as if the latter were no tougher in substance than cardboard. Passing to the stone-masons' yards and shops, work of various kinds and materials was in progress, both by hand and steam power, and machinery appeared to be used in this department of labour almost as largely as in wood-working. . . . Particular attention was attracted to a heavy lathe in which a workman was turning a large baluster in Portland stone. The speed at which the work revolved was necessarily slow, but still it was great enough to cause chips large and small, and a cloud of coarse and fine dust, to fly off under the touch of the tool or chisel (somewhat resembling a short but rather stout crowbar) directly in the face of the workman, who was, however, unprovided with a mask or respirator of any kind. When we state that the chips and dust were thrown a distance of three or four yards, and that the workman was veritably in the thick of it all, the extremely injurious effect of the particles of stone inhaled by the lungs of the mason must be obvious. We spoke to the workman, who appeared to be a very intelligent man, and he said he formerly wore a respirator, but he found it too "choky," from the fact, we suppose, that it became clogged with the particles of stone; he had, therefore, discontinued it, his own remark being that "one can but be choked, either way." . . . The visitors passed from the stonemason's department to that of the marble-masons, where machinery is also largely made use of. . . . In this part of the establishment is another museum, where, in addition to a large number of beautiful chimneypieces, specimens of every variety of marble used for building purposes are exhibited, one of the rarest and most beautiful being that known as "Chinese." The visitors next proceeded to the iron foundry, where the operations of moulding and casting were practically exemplified, under the superintendence of Mr. Ridge, the foundry foreman. The engineers' shop was next visited, and it will convey some idea of the resources of the firm when we say that here are made all the machinery and mechanical appliances of every kind used on the establishment. Next was visited the stove-makers' and locksmiths' shops, the visitors passing thence to the terra-cotta and cement works and kilns. . . . One of the kilns, only opened the same morning, was full of very well finished terra-cotta airbricks; but all kinds of terra-cotta work, up to the largest and most elaborately moulded pieces, such as vases, cornices, &c., are made on the premises, and a very fine collection of specimens was exhibited in an adjoining show-room. The terra-cotta is compounded of broken glass and crockery (a quantity of which was being ground to powder for the purpose by two pairs of edge-runners), and London or Poole clay, according to the colour and material desired. In this department, too, the firm makes all its own cements. The visitors having passed through the plasterers' and modelling shops, where some interesting work was to be seen in progress, concluded their tour of inspection at half-past five by a vist to the general show rooms. . . . About 800 men are employed on the premises; but reckoning those engaged on "works in progress," the number of men employed by the firm is about 3,000. That machinery is very largely employed will already have been gathered from what is previously stated; but the fact that upwards of a mile of shafting is used will perhaps convey this more strikingly. . . . The firm prefer to manufacture their own goods and materials throughout, not only from motives of economy, but from the higher consideration of securing the best of quality.

William Cubitt was not an innovator, nor did he change the face of large parts of London as his brother did, but he was probably better known and more widely known than him. He became a leading figure in the City of London. Sheriff of London and Middlesex in 1847, he was elected alderman of the City four years later, Lord Mayor in 1860 and, thanks to the hospitality he offered, again Lord Mayor in the following year. He was a well-known personality. A story was told of him as Lord Mayor trying a case at the Mansion House and asking a witness to make sure his evidence was correct. 'When I arose this morning', said Cubitt, 'I could have sworn that I put my watch into my pocket and I have only just missed it, and now I recollect that I left it on my dressing table.' When he returned home he was surprised that his wife asked why he had sent so many messengers, one after the other, for his watch and chain. During these years, sixteen in all, with one short break, he was also Member of Parliament for Andover, Wiltshire. He spoke very seldom in the House, and then only when he knew what he was talking about. For instance he made a short, well-argued speech in May 1850 in a debate about the rising cost of building the new Houses of Parliament. He pointed out that it was very difficult to estimate accurately when there was no experience of such an enormous building, that the drawings which had been submitted to the House had given no idea of the decorations added later, and that he could not explain how the building had come to be more decorated than was intended. 'But', he went on, 'the country did well to erect a structure to last for centuries, in the highest style of art which could possibly be produced; no building in Europe, ancient or modern, would compare with the new Palace when finished; a country which was spending 50 to 60 million pounds per annum should not grudge £200,000 to £300,000 for ten or twelve years for such a building' (Hear, hear).

Although a conservative he was a liberal kind of conservative. He supported free trade and abolition of the duty on malt. As an employer who paid the wages of thousands of workers he saw the importance of cheap food imports and beer.

He worked hard for many good causes. By 1844 his reputation was such that he was appointed a member of the Royal Commission on the State of Large Towns and Populous Districts, which included, among others, the Duke of Buccleuch and Robert Stephenson. This enquired into the appalling lack of sanitation and contaminated water supply, throughout the country, and it led to the first Public Health Act of 1848. Cubitt was also president of St. Bartholomew's Hospital. But he became best known at the time of the Lancashire cotton famine.

When the American Civil War broke out in April 1861, England depended on America for four-fifths of its supply of raw cotton, and about a fifth of its population made a living from the cotton trade. By the end of that year supplies had stopped. As a result there was great unemployment and hardship in Lancashire and the neighbouring counties. Half a million peple were living on poor law relief and charity. Cubitt, as treasurer of the Mansion House relief fund in London was very successful in raising money. Through him many thousands of pounds were sent to local relief committees in Lancashire very promptly and well before any official action was taken. At Christmas 'the dole of 8d. per head, in the proportion of 6d. in flesh meat and 2d. in groceries, afforded the poor a substantial dinner of flesh meat to the whole family for three or four days; a pleasant and wholesome change, for the time from the low oatmeal diet to which so many had been reduced.'

When Cubitt died there were daughters, two of whom married

the sons of a fellow alderman, to inherit his wealth. His only son died while a student at Cambridge. On the Sunday after his death, 8th November 1863, the church bells rang a muffled peal in his honour in fifty-four towns and villages in Lancashire. At one place, Ashton-under-Lyne, there was a special demonstration of gratitude. The vicar reported that the workers 'assembled in committees and resolved to ask a funeral sermon to be preached in his memory; and for this, that they would go to the church in a body, on the occasion, in funeral procession as mourners. And so the men having formed their procession at half past 5 o'clock, walked up in long train, two abreast, of mourners dressed in as decent black as they could procure, and many of them carrying lighted torches as they moved through the drizzling wet of a dark and boisterous evening, they took their part heartily in the long service in an inconveniently crammed church, where many had to stand the whole time. They had begged the use of black cloth from one of the clothiers of the town with which they draped the pulpit'.

The third of the Cubitts, Lewis, was, unlike his brothers, a trained architect. This Cubitt was a man of the railway age which really started with the Liverpool and Manchester Railway of 1831. As a young man he worked for his brothers' firm, designing some of their houses in Bloomsbury and Belgravia, but then he became a designer of railway stations in that first era of railway construction. The first two stations were at Dover on the South Eastern Railway and at Bricklayers Arms off the Old Kent Road, London, for the South Eastern and the Croydon Railways. The station for which his name is still well known is King's Cross built in 1850–1852 for the Great Northern Railway which was just in time to bring the crowds to the Great Exhibition. The two tremendous semi-circular arches on the plain brick front are simply an extension of the great shed roofs inside, which were an innovation at the time. A few years later a Victorian writer said: 'The Great Northern Terminus is not graceful, but it is simple, characteristic and true. No one would mistake its nature and use.' St. Pancras built almost next door and only twenty years later by Scott (see Chapter 3) is a complete contrast.

THE BUILDING WORKERS

Well before the industrial revolution got under way and even longer before big building contractors like the Cubitts appeared on the scene, the building craftsmen like most other skilled trades had their clubs and friendly societies. But they did not form trade unions until some time after the workers in other industries, in textiles, engineering, coal, mining. This was for two reasons. Their work had not been changed by machinery or by the factories, and they were not yet employed by big capitalists. A man who wanted to build a house still employed a master craftsman in each trade who in turn employed journeymen and apprentices.

The clubs and friendly societies giving sick pay and funeral expenses, sprang up all over England. They were local, each based on a particular town or village. Some of them were also limited to a particular trade or craft, and many of these developed later into trade unions. An example of the early clubs is the Preston Joiners. Here are the first entries in its cash book:

> '1801—Feb. 9   By 1 Quire of paper   1s. 6d.
>   By Ale 4 Glasses   8d.
>   By 2 books   8d.
>   Feb. 25   By Expenses of Court Meeting 8s. 8d.
>   27   By 8 Glasses of Ale   1s. 4d.'

The club grew and opened lodges at Bolton and Blackburn. Its money came from subscriptions and apprentices' fees. As well as spending on beer and meetings the club paid the expenses of its members tramping from town to town in search of work.

> '1808—Sept 8th Paid John Holliday for tramp money   3s. 6d.'

Although at first the clubs were concerned with keeping outsiders from poaching on their work, gradually they came to discuss wages and sometimes to combine against employers. As early as 1720, during Wren's lifetime, in London 'the Journeymen Carpenters, Bricklayers and Joyners have taken some steps for that purpose, and only wait to see the events of others'. Thirty years later in London there was a 'club of journeymen painters that would not work nor let others work'. And in 1789 four carpenters were charged at the Old Bailey with 'conspiring against their masters to raise wages'. They were bound over, after expressing their sorrow for having combined their fellow workers against the employers.

1789 was the year of the French Revolution. When the artisans of England took hold of the ideas of liberty, equality and fraternity they caused great alarm among the ruling classes. Corresponding Societies sprang up to spread the ideas from France. The Corresponding Society in London, which rose to about 10,000 members, included carpenters and bricklayers, among them, as well as many other trades. At the same time the spread of trade unions among the engineering workers, weavers and miners, caused the Government to pass the Combination Acts of 1799 and 1800. These made it a criminal offence for any worker to combine with another to get an increase in wages or decrease in hours.

Trade unionists were widely prosecuted and the unions were driven underground. However, the building workers' small local clubs and societies survived. The painters, for instance, accepted the situation. Very soon after the Acts their journeymen and masters met and agreed:

'That fair equitable and liberal wages as between Master and Journeyman should be paid, namely, at the rate of one guinea per week for good and able workmen—a day's work being reckoned from 6 o'clock in the morning till 6 o'clock in the evening—and inferior workmen according to their abilities.

That the Act to prevent unlawful combinations of workmen be enforced.'

All the same, many of the societies acted as trade unions as well as providing security benefits. One of them, the Friendly Society of Carpenters and Joiners which met at the Running Horse, London, in 1800 explained how a member was protected: 'The great quantity of expensive tools which are necessary for him to pursue his daily avocation expose him to continual liability of losses from fire and robbery and as such losses cannot possibly be at all times avoided it becomes necessary to take preservative measures against unjust innovations.' 'Innovations' meant changes in the working rules, made by the employer.

In spite of the law the workers continued to combine as they could, whether in secret or in the disguise of friendly societies. In Leeds only two years after the Act the mayor complained to the lord lieutenant of the West Riding:

'Perquisites, privileges, time, mode of labour, rate, who shall be employed etc., etc.,—all are now dependant upon the fiats of our workmen, beyond all appeal; and all branches are struggling for their share of these new powers. It is now a confirmed thing that a bricklayer, mason, carpenter, wheelwright, etc., shall have 3s. per week higher wages in Leeds or in Manchester than at Wakefield, York, Hull or Rochdale.'

These were years of rapid expansion in the building industry. Eight years later in London, the bricklayers, carpenters and plasterers were able to strike and get an increase in wages from 28s. to 30s. a week.

These were also the years of the long war with France and rising prices. The carpenters in London were better organised than some. They had five societies there with about 2,500 members and considerable funds. A stocking-weaver on a deputation to London in 1812 met them there and wrote to his friends:

'We have engaged the same Room, where the carpenter committee sat. . . . We have had an opportunity of speaking to them on the subject, they thought we possessed a fund on a permanent principle to answer any demand, at any time, and if that had been the case would have lent us two or three thousand pounds (for there is £20,000 in the fund belonging to that Trade) but When they understood our Trade kept no regular fund to support itself, instead of lending us money, Their noses underwent a Mechanical turn upwards, and each saluted the other with a significant stare, Ejaculating, Lord bless us!!! what fools!!!!, they richly deserve all they put! and ten times more!!! . . . What could our Trade be, if we did not combine together! perhaps as poor as you are, at this day! Look at other Trades! They all Combine (the Spitalfields weavers excepted, and what a Miserable Condition are they in). See the Tailors, Shoemakers, Bookbinders, Gold beaters, Printers, Bricklayers, Coatmakers, Hatters, Curriers, Masons, Whitesmiths, none of these Trades Receive Less than 30s. a week, and from that to five guineas this is all done by Combination. . . .'

Soon, although the expansion of building was slowing down, the London carpenters pushed their wages up to 35s. a week, but with the slump in building after the war they were faced with a cut. They claimed that 'single men only can live and this accounts for the present wretched state of society—no wonder the streets are full of prostitutes when matrimony is the sure road to poverty'. They struck but under the Combination Acts their leaders were sentenced to prison, two for a year and three for one month. They prepared a case against the employers who were also liable under the Acts but 'they did not prove it because the magistrates quashed the information and they could not prove it.'

When the Combination Acts were repealed in 1824 and more prosperous years came the trade unions flourished for a short while. 'In Staffordshire the carpenters were the first to strike, and then every other trade turned out in rotation' according to a newspaper report in the following year. In London, the rebuilding of Buckingham Palace under John Nash was stopped by a strike of carpenters which was ended when the Coldstream Guards were called out. The trade unions very soon lost most of their gains but the carpenters took a big step forward. At a delegate meeting in London on 19th July 1827, they formed the first national union in the building industry, the Friendly Society of Operative House Carpenters and Joiners of Great Britain. The preamble to their rules said:

'We consider it absolutely necessary that a firm compact of interests should exist among the operative House Carpenters and Joiners of Great Britain and Ireland. We therefore as representatives of the several Lodges of the trade, in our name and in the name of all who may adhere to us, unite in the bonds of friendship for the amelioration of the evils besetting our trade; the advancement of the rights and privileges of labour, the cultivation of brotherly affection and mutual regard for each other's welfare.'

This society, soon renamed the General Union, lasted until 1920 when it was absorbed in the Amalgamated Society of Woodworkers. Two years after the carpenters, the bricklayers also formed their national society. Thus the way was prepared for an even bigger and more ambitious union several years later. This was brought about by 'the evils besetting our trade'. What were they?

Some had existed for a long time, for instance the laying off during the winter months. In the eighteenth century the London building workers worked a season of only seven months in the year. At the time of the above events many of the bricklayers, plasterers, masons, etc., had work for only nine months. Other evils were typical of the times. To escape from this situation, and from being at the mercy of the sub-contractor, many craftsmen tried speculative building. In the squalid new towns of the industrial revolution like Manchester there was plenty of opportunity.

'These towns ... have been erected with the utmost disregard of everything except the immediate advantage of the speculative builder. A carpenter and builder unite to buy a series of building sites and cover them with so-called houses ... in general the streets are unpaved, with a dungheap or ditch in the middle; the houses are built back to back without ventilation or drainage, and whole families are limited to a corner of a cellar or garret.'

Even so, a great many went bankrupt; 'the land rented in hope, materials secured on credit, a mortgage raised on the half built house before it is sold or leased, and a high risk of bankruptcy'.

Building workers were urgently needed in the new towns and for them at times, there was plenty of work and good money. Their numbers increased to about 400,000 in 1831, the third largest occupation in the country, after agriculture and domestic service. But in the rush of jerry-building standards of work fell. Apprenticeship regulations had been abolished by Parliament and any labourer could pick up enough knowlege to build the shoddy houses. The masons complained about 'the system which existed till of late, allowing anybody to learn our trade and serve what time they pleased'. But when the booms broke wages collapsed. This was a basic 'evil', new to many; that the growth of industry brought booms and slumps, stop and go, to workers who had known a more steady life. It had also brought the general contractor who took the place of the master craftsmen each of whom contracted for his own part of the job. Thomas Cubitt had started his large-scale developments.

One of the slumps in building started in 1830. Two years later the building workers formed the first national union for all trades, the Operative Builders' Union. It so happened that another kind of 'evil' occurred at that time: it was the feeling of the working class that it had been cheated by the Reform Act of 1832. The campaign for reform of Parliament and for the vote had been a violent one. The employers had led out the workers to demonstrate for it but it was the employers who benefited from the Act. Realisation of this fact meant that the Operative Builders' Union (O.B.U.) had much wider and ambitious aims than the ordinary ones of a trade union.

The O.B.U. was a federation of the existing unions of all the different trades. Within a year the membership had risen to about 40,000, an astonishing number to contemporaries. The bricklayers and masons were the biggest elements, although the painters numbered 6,000 to 7,000.

The union was well organised, in seven sections: stonemasons, painters, plasterers, plumbers and glaziers, carpenters, bricklayers, and slaters. The basis of each section was the local lodge, limited to members of one craft. The stonemasons' rules stated, 'No other member to visit another Lodge that is not the same trade unless he is parti-

cularly requested. And then he shall ask the President of such Lodge
as he wishes to visit and to be upstanding during his discourse. And that
he shall withdraw as soon as he has done'. Above the local lodge was
the district lodge. For instance, the stonemasons had eleven districts;
in London, Leicester, Nottingham, Cheltenham, Birmingham,
Potteries, Chester, Lancashire, Yorkshire, Kendal, Newcastle, and 100
lodges in these districts. At district level of the O.B.U. there was a
district Grand Master to co-ordinate the seven sections or trades.
Above the districts was the central authority of the O.B.U., the Grand
Lodge, which was sometimes called the Builders' Parliament. It was
composed of one delegate from every lodge; the Builders' Parliament
in September 1833 at Manchester was attended by about 500 delegates
and sat for a full week. Meeting three times a year, it decided policy
and controlled finance, and elected the Grand Officers. The union's
day-to-day business was directed by a Grand Committee of the Grand
Lodge, which was composed of the Secretary and two committee
members of each section, plus the three Grand (National) Officers.
The Committee controlled strikes for increases in wages, for if it
approved them it collected levies for them from the lodges. Strikes
against reductions in wages did not require approval by the Grand
Committee and lodges were expected to find their own money.

'The object of this society', ran the rules, 'shall be to advance and equal-
ize the price of Labour in every Branch of the trade we admit into
this society'. Every member had to take an oath of loyalty:

'I do before Almighty God and this Loyal Lodge most solemnly swear
that I will not work for any master that is not in the Union nor will I
work with any illegal man or men but will do my best for the support
of wages, and most solemnly swear to keep inviolate all the secrets of
this Order, nor will I write or cause to be wrote, print, mark, either
in Stone, Marble, Brass, Paper or Sand anything connected with this
Order, so keep me God and keep me steadfast in this my obligation;
and I further promise to do my best to bring all legal men that I am
connected with into this Order; and if ever I reveal any of the rules
may what is before me plunge my soul into eternity.'

'What is before me' was a sword pointed at the member.
    The aims of the O.B.U. became wider than 'to advance the price of
labour', important though this was. It came under the influence of
Robert Owen, the successful manufacturer turned socialist, who said

that the answer to the evils of the time was to replace competitive capitalism by co-operative production and so change the whole of society. This became a rallying cry for many thousands. Owen had already persuaded other trade unions to run their own co-operative workshops and he had no difficulty in persuading the O.B.U. to change itself into a Builders' Guild which would, before long, take over the whole industry. Owen described the meeting:

'A large number came on account of the information which was conveyed to the different lodges that our new plans of society would be submitted to them for consideration. Many of them were quite novices in the new doctrine. The assembly might be said to consist of three classes—first, of those who were entirely unacquainted with the new doctrine—second, of those who were partially instructed in its principles—and third, of those who were comparatively well informed.'

This 'Grand National Guild of Builders' was to cut out the general contractors by working directly for the customer. There were to be masters, journeymen and apprentices but the masters would be elected in the lodges. These lodges were to consist of architects as well as all the different kinds of craftsmen, 'and also quarriers, brickmakers and labourers as soon as they can be prepared with better habits and more knowledge to enable them to act for themselves, assisted by the other branches who will have an overwhelming interest to improve the mind, morals, and general condition of their families in the shortest time.'

The Guild would have its own bank which would issue its own paper money. The O.B.U. announced a 'friendly declaration' to all the British dominions. Its far-reaching aims can be seen from the following extracts:

'Our eyes have been opened upon this subject, we have discovered that our natural and acquired resources are unlimitable and almost inexhaustible and that we and all the industrious classes, have been made the victims of the most lamentable and grievous errors by those who have directed the producing powers of the country. That in consequence we have been kept in ignorance when we might have been made intelligent—reduced to poverty when we might have been made to superabound in riches—divided in our sentiments, feelings and interests when we might have been united in each of them—degraded to the lowest scale in language, habits, conditions and public estimation, making us despised and oppressed by all, when we might have been placed in a situation to be highly esteemed, and respected by every other portion of the human race, and when, also, we might have been made far more valuable to our own country and to the population of the world than we have every yet been or can become while the present errors in directing the resources of the country shall be continued.

'It is now evident to us that those who have hitherto advised the Authorities of these realms in devising the Institutions of our country were themselves ignorant of the first principles requisite to be known to establish and maintain a prosperous and superior state of society. With this view we have formed ourselves into a National Building Guild of Brothers, to enable us to erect buildings of every description upon the most extensive scale in England, Scotland and Ireland.

'1st.—We shall be enabled to erect all manner of dwellings and other architectural designs for the public more expeditiously, substantially and economically than any Masters can build them under the individual system of competitition.

'2nd.—We shall be enabled to withdraw all our Brethren of the National Builders Guild and their Families from being a burden upon the public, for they will be supported in old age, infancy, sickness or infirmity of any kind from the general funds of the Guild.

'3rd.—None of the Brethren will be unemployed when they desire to work, for when the public do not require their services they will be employed by the Guild to erect superior dwellings and other buildings for themselves, under superior arrangements, that they, their wives and their children may live continuously surrounded by those virtuous external circumstances which alone can form an intelligent, prosperous good and happy population.

'4th.—We shall be enabled to determine upon a just and equitable remuneration or wages for the services of the Brethren according to their skill and conduct when employed by the public.

'5th.—We shall also be placed in a position to decide upon the amount of work or service to be performed, each day, by the Brethren, in order that none may be oppressed by labour beyond their powers of body or mind.

'6th.—We shall be enabled to form arrangements in all parts of the British dominions to re-educate all our adult Brethren that they may enjoy a superior mode of existence, by acquiring new and better dispositions, habits, manners language and conduct, in order that they may become such examples for their children as are requisite to do justice to all young persons whose characters are to be formed to become good practical members of society.

'9th.—We will exhibit to the world, in a plain and simple manner, by our quiet example, how easily the most valuable wealth may be produced in superfluity beyond the wants of the population of all countries; also how beneficially, for the Producing Classes (and all classes will soon perceive their interest on becoming superior pro-ducers) the present artificial, inaccurate and therefore injurious circu-lating medium for the exchange of our riches, may be superseded by an equitable, accurate and therefore rational representation of real wealth, and as a consequence of these important advances in true civilization, how beautifully, yet how accurately the causes which generate the bad passions and all the vices and corruptions atributed to human nature, shall gradually diminish until they all die a natural death and be known no more, except as matter of past history, and thus by con-trast, be the cause for everlasting rejoicing.

'10th.—We shall by these and other means now easy of adopting speedily open the road to remove the causes of individual and national competition, or individual and national contests, jealousies and wars, and enable all to discover their true individual interests and thereby establish peace, goodwill and harmony, not only among the Brethren of the Building Guild, but also by their example among the human race for ever.

'11th.—We shall secure to the present Masters of all the Building Branches who well understand their business a far more advantageous and secure position in Society than they have or can have under the system of individual competition between Master and Master and Man and Man, and we shall open the way to unite their interests cordially, firmly and permanently with the real body of the National Builders Guild.'

In order to carry out these plans the Union started to build its own premises, a great Guildhall in Birmingham at a cost of £1,000, which was celebrated as 'the beginning of a new era in the condition of the whole of the working classes of the world'.

Yet after only two years, by the end of 1834, the O.B.U. and the Guild had fallen apart, and the Guildhall had been sold unfinished and turned into a warehouse. How did this happen?

At first the Union was successful in its demands. These were mainly that there should be no general contracting, and therefore had the support of many small masters. In Manchester placards demanded 'that no new building should be erected by contract with one person'. But the masters became alarmed and combined to break the Union. One of them received the following letter:

'Mr. Goodess,

I have to inform you that we have been informed of your nefarious proceedings, (N.B.) your proposition that if the masters of all the Building Trades will turn out their men for one fortnight they will overthrow the Union at once; now I have to inform you that if this be your return for us striking the shackles off your legs from the contractors, you cannot speak ever so privatly but we hear of it as soon as it comes from your mouth and if you can contradict this statement you will oblidge the operative Plasters and Painters.'

In Lancashire open conflict broke out. The Union thought that the masters were breaking their promises about general contracting but the masters found they could not avoid the system. They presented the Union with the Document, which all workers were required to sign:

'We, the undersigned . . . do hereby declare that we are not in any way connected with the General Union of the Building Trades and that we do not and will not contribute to the support of such members of the said union as are or may be out of work in consequence of belonging to such union.

June 15th, 1822.

The reply was a solid strike.

At the same time in Birmingham one of the largest contractors sacked every man who belonged to the Union. He was written to as follows:

'Sir,—We the delegates of the several Lodges of the Building Trades elected for the purpose of correcting the abuses which have crept into the modes of undertaking and transacting business, do hereby give you notice that you will receive no assistance from the working men in any of our bodies to enable you to fulfil an engagement which we understand you have entered into with the Governors of the Free Grammar School to erect a New School in New Street, unless you comply with the following conditions.

Aware that it is our labour alone that can carry into effect what you have undertaken, we cannot but view ourselves as parties to your engagement, if that engagement is ever fulfilled; and as you had no authority from us to make such an engagement, nor had you any legitimate right to barter our labour at prices fixed by yourself, we call upon you to exhibit to our several Lodges your detailed estimates of quantities and prices at which you have taken the work, and we call upon you to arrange with us a fixed percentage of profit for your own services in conducting the building and in finding the material on which our labour is to be applied.

'Should we find upon examination that you have fixed equitable prices which will not only remunerate you for your superintendence but us for our toil, we have no objection upon a clear understanding to become parties to the contract and will see you through it, after your having entered yourself a member of our body, and after your having been duly *elected* to occupy that office you have *assumed*.'

During these months there were also strikes or lock-outs at Leeds, Worcester and Nottingham.

In the meantime the Union members began to quarrel among themselves. A small minority called the Exclusives, mainly carpenters, had never agreed to the big union and its wide socialist aims. They started to demand that the Union should be broken up into its component societies. The *Pioneer*, the Union's weekly journal, said, 'We will give them a new name, we will call them the Pukes—it is a sickening idea—and will remind us that we are looking upon something that is filthy'.

The Union suffered its first defeat. After a strike of sixteen weeks the Lancashire men had to give in for lack of funds and signed the document. In Birmingham they had failed to stop the school contract and instead pressed on with the costly Guildhall. But money was short and work stopped in February 1834 and restarted only in June for a time. In March 1834 a fatal blow was struck at the whole trade union movement when the labourers at Tolpuddle were sentenced to seven years transportation for administering unlawful oaths when enrolling members. Then in July that year the employers in London declared a lock-out. This dispute started at Cubitts' yards. The workers employed there refused to drink the beer which was usually brought in, because the brewers had refused to employ trade unionists. Cubitts' would not allow any other beer to be drunk on their premises and when the workers defied this ban they were locked out.* The other employers,

* Probably this was not the only cause of the dispute. In February the labourers employed by Cubitts' on the new Fishmongers' Hall in Lower Thames Street

taking the opportunity to attack the O.B.U. and its resistance to piece-work and general contracting, demanded that the workers should sign the 'document'. This started a struggle which dragged on for four months and ended in defeat for the workers.

All these setbacks encouraged the Exclusives. First the masons, carried a resolution at the Grand Lodge in September 1834 'that this society do come under Exclusive Government' and left the O.B.U. Then the other trades followed so that by the end of that year the Operative Builders Union had fallen apart. The dream of workers' unity and a new society was over for the time being.

FURTHER READING:

J. Davis, 'John Nash', *Country Life*, 1966.
J. Summerson, *John Nash*, Allen & Unwin 1949.
H. Hobhouse, *Thomas Cubitt–Master Builder*, Macmillan 1971.
Sir Stephen Tallents, *Man and Boy*, Faber 1943.

---

had struck for an increase in wages. They were joined by the bricklayers, number-ing altogether 200, when they also demanded that the foreman be sacked.

2   The Palace of Westminster

The biggest and most elaborate building of stone constructed in Britain during the nineteenth century was the Palace of Westminster. It took nearly thirty years of continuous work to complete. No wonder the stonemasons had a powerful and stable trade union throughout the whole time. The new Houses of Parliament occupied a site of about eight acres. On its lowest front, on the river Thames, it was 940 feet in length, and its greatest width about 340 feet. The enormous building aroused tremendous controversy but it came to be admired and wondered at by the whole country. Started soon after the Reform Act of 1832 it was regarded with pride as the home of democracy and as the mother of parliaments in other countries. The man who designed it was a Londoner called Charles Barry. Like the architect of that other great building a hundred and fifty years earlier, St. Paul's Cathedral, he controlled construction practically throughout but, unlike Wren, he died before the work was finished. England had changed fundamentally in that hundred and fifty years. There is a great contrast between the appearance and style of the two buildings; and also between the two men.

SIR CHARLES BARRY (1795–1860)

Barry was born in Westminster in a house in Bridge Street opposite to where the tower of Big Ben was to stand. His father was a prosperous stationer who supplied the Government stationery office. His mother died when he was three and he and his three elder brothers were brought up by his stepmother. His father died when Barry was only ten so that the stepmother had to look after the family business as well as care for the children. The boy was sent to three schools but none of them were much good. At the second school it was said 'the master paid little attention to it, being very dissolute and absenting himself for weeks together'. When he left school finally at the age of fifteen, he had learnt only English, arithmetic and handwriting.

Already he had shown a marked ability for drawing. He was,

therefore, articled to a firm of architects and surveyors, Middleton and Bailey, of Paradise Row, Lambeth. There he received a good professional training for six years. The firm's work was mainly in surveying and he learnt a good deal about the practical side of the business, measuring, calculating, pricing and valuing, but little about design. This he had to teach himself, and did it very successfully. For, from the age of seventeen, each year until he was twenty, his architectural drawings were exhibited at the Royal Academy. The first one was a drawing of the interior of Westminster Hall, the eleventh-century building which was to be the nucleus of the new Houses of Parliament. His other designs were of a church, a museum and library, and a 'nobleman's mansion'.

At home he expressed his personality, or his need, in another way. He had a small attic bedroom and he made it to look like a hermitage, a cave with openings looking out on a sunny landscape, with wood, canvas and paint. He was exercising his ingenuity and imagination; later on he could never enter a room without seeing its possibilities for conversion. It was a way of developing his talents and also of expressing his needs. The atmosphere at home was happy, quiet and encouraging.

The year 1815 brought peace to Europe and after twelve years Europe was open to the English middle class. Many sallied across the channel that year. Barry completed his articles and on reaching the age of twenty-one inherited a few hundred pounds from his father. Instead of staying with his firm he decided to take a chance and spend the money, which was all he had, on a tour to study the architecture of the continent. He left in 1817 and did not return until 1820. Just before going he became engaged to the daughter of another stationer, whom he married five years later.

The three years of travel were very important for Barry. He learned about all kinds of building in many countries and when he returned he knew what he wanted to do as an architect and what sort of building he would design. He visited not only the usual places in France, Italy and Greece but also the much less known parts of Egypt and Syria. For the first nine months he lived mainly in Paris, Rome and Milan. In Paris the sharp contrast between the dirt, bad drainage and bad paving of the city and the showy life of the Bourbon régime made a great impression. Being interested in art as much as architecture he spent many days in the Louvre and in the Vatican. In his notebooks he recorded precise measurements of the buildings he examined. This led him into trouble at a small town in Italy, then under the Austrian heel, where he was sketching. Soldiers entered his bedroom and

ordered him to go; when he refused and pointed a pistol they put a
sentinel at his door until he left.

At Rome he seized the opportunity to travel farther to Greece and
Turkey with three men who were already well known in architecture
and art. From Naples and Pompeii they rode across Apulia to Bari
where they were unable to find a ship for Corfu and crossed the
150 miles of sea in a small coasting vessel, a felucca. Having spent a
month in Athens they left in June 1818 via the Cyclades, Delos, and
Smyrna for Constantinople. There a wealthy young Scot offered to
take Barry with him paying him £200 a year in return for all the
sketches of buildings and scenery he would make. Barry agreed as he
could keep copies of his sketches. In Egypt, where they spent four
months, they travelled some 1,500 miles, ascending the Nile as far as
the second cataract near Wadi Halfa. The ancient Egyptian buildings
and tombs thousands of years old, made a deep impression on Barry.
In November they came in sight of the great temple of Denderu: he
wrote in his journal:

'It astonished us by its unexpected magnitude, and gave me a high
idea of the skill and knowledge of the principles of architecture dis-
played by the Egyptians. There is something so unique and striking in
its grand features, and such endless labour and ingenuity in its ornaments and hieroglyphics, that it opens to one an entirely new field.
No objects I have yet seen, not even the Parthenon itself (the truest
model of beauty and symmetry existing), has made so forcible an
impression upon me. The most striking feature of the building is its
vast portico, six columns in front and four in depth, giving a depth of
shadow and an air of majestic gravity such as I have never before seen.'

At Aswan they transferred from their large river vessel to four small
boats and went on another 150 miles or so to the second cataract.

He started home in June 1819, leaving 500 drawings with his patron;
but evidently in no hurry took more than a year on the way. He arrived
in Rome in January 1820, after staying in Cyprus, Malta and Sicily.
There he made a life-long friend: 'In my first interview with him', he
wrote, 'I saw immediately that he was a man with whom I could
coalesce and become intimate; and the result is that I now reckon him
among the few sincere friends that one can hope to obtain in the world.'
His drawings made him well known and he met the English aristocracy
living in Rome. The return tour of Italy was important for Barry.
A great admirer of Inigo Jones, he hunted out every building by Jones's

master, Palladio. The Italian palaces and villas which he examined so
carefully had a strong influence on his own work later in England.

Back in London, Barry started to earn his living as an architect. He
was twenty-five with very little money; he took a small house in Ely
Place, Holborn, and began to enter public competitions. At first he had
no success and thought of emigrating. Yet in fifteen years he acquired
such skill and experience that he won the competition for the biggest
job of all, the Houses of Parliament.

Fortunately for Barry many new churches were being built. In many
of the new towns of the industrial revolution there were no churches
at all. Parliament granted one and a half million pounds to build them,
and another four and a half million came from subscribers. At a time
of political unrest and danger the authorities thought it wise to provide
churches for the working class. Unfortunately for Barry, who had
trained himself in the classical style of horizontal lines and regular
blocks, the style of the new churches was a revival of the old medieval
Gothic style of vertical lines, spires and pointed arches. However, he
set to and learned this latest style and was quickly able to take advantage
of the boom in churches. His first church, at Prestwich, was begun only
two years after his return to London. In addition, the influential
friends he had in Rome introduced him into aristocratic circles and
brought him useful social contacts.

Once started, he made rapid progress, designing other churches in
Manchester, Oldham, Brighton and Islington. In Islington his three
churches cost £34,000, a large sum for those days, of which he
received his 5% fee. He was at last able to marry his patient fiancée and
a few years later to move to a more fashionable address near Cavendish
Square in the West End.

The building that really made his name, when he was thirty-five,
was the Travellers' Club in Pall Mall. Again he was lucky enough to
be in an era of club making and club-house building in London by the
wealthy landowning and middle classes. Out of four such handsome
buildings erected in Pall Mall between 1828 and 1837, Barry designed
two: the Travellers' and the Reform Club, formed 'to promote the
social intercourse of the Reformers of the United Kingdom'. Both of
them he designed in the style of the palaces which he had seen in
Italy; the first was modelled on the Pandolfini Palace at Florence, the
second on the Farnese Palace at Rome. This was the style he really
preferred. At the same time he kept his hand in on the increasingly
fashionable but quite different style, the new Gothic, and designed
King Edward VI's School, Birmingham, in that manner. (It was there

that the operative Builders' Union tried unsuccessfully, to control the contractor.)

On the night of 16th October 1834, Barry was returning from Brighton on the coach when he saw the glare of a great fire in London. The Houses of Parliament were burning down. Someone had wrongly burnt old tally sticks in the stoves under the House of Lords and defective flues had done the rest. Great crowds gathered round the spectacle, many watched from barges and small boats in the Thames. The crowd cheered as the roofs fell in, with a specially loud cheer for the collapse of the Lords' library. Barry saw his opportunity. The design for the new building was thrown open to competition in which four premiums of £500 each were to be awarded, the first premium going to the accepted design. Ninety-seven competitors submitted designs; against these, in February 1836, Barry won the first premium. It had taken six months' continuous work to prepare the drawings for the great project. Because of the surrounding buildings it was laid down that the style had to be either Elizabethan or Gothic. Barry chose the Gothic although he would have much preferred to design in the Italian style as with his Pall Mall clubs, and if he had been allowed to, he would have followed Inigo Jones's grand design for the Palace at Whitehall. The commissioners who were responsible for selecting Barry's design gave their reasons. They explained that sound or audibility and ventilation had not counted for much in their decision because not enough was known about those subjects. What had decided them was 'the beauty and grandeur of the general design, its practicability, the skill shown in the various arrangements of the building, and the accommodation afforded; the attention paid to the instructions delivered, as well as the equal distribution of light and air through every part of the structure'.

When Parliament agreed to the plan for its new home, against some opposition, the time for completion was fixed at six years, and the cost approved was approximately £700,000. In fact it took nearly five times that period and about three times the cost. If Barry had known this his pleasure in his triumph would have been much less. As it was, at the age of forty, he had reached the top in his profession.

The first stone of the great new building was laid at the south-east angle of the Speaker's house, without any ceremony, by Barry's wife. This was not until April 1840. It took more than a year to decide what stone should be used. Eventually the experts chose magnesian limestone from Bolsover Moor, Derbyshire; later on it was found there was not enough at Bolsover and stone from Anston, Yorkshire, had to be used.

But the delay at the beginning arose mainly from the nature of the site right on the bank of the river Thames. In fact, because the river front was not parallel to the line of old Westminster Hall, it had to be shifted by advancing the embankment into the river near Westminster Bridge and setting it back at the other end. The river wall was begun in July 1837 but it was not completed until 1839. For this a well-known civil engineer shared the responsibility with Barry. The main problem was that the soil of the river bed and part of the soil where the building was to go, was very treacherous. Barry had the foundations of the wall laid on concrete, twelve feet thick in some places, behind a coffer dam. He left in position the piles of the coffer dam cut off at river bed level, so as to give extra protection to the wall against the scouring action of the river. Then the wall was faced with Aberdeen granite, the terrace laid and the main foundations brought up to that level.

The whole of the vast complicated construction was carried out by a series of large contracts. These were awarded by the Government department concerned (the Office of Woods, etc.) as a result of competitive tender or recommendation by Barry. The first contract, dated 11th September, 1837, for £74,373, was with Henry and John Lee for construction of the coffer dam and river wall. There was a penalty clause of £10 a day. The coffer dam was to be of 'good and suitable Memel or Danzig fir timber'. The contract stated:

'. . . the space between the river wall and the front wall of the building is to be brought up to the level of the top of the walls with concrete composed of 10 measures of gravel and sand to one of unslaked lime washed in with water.
'. . . the mortar throughout is to be comprised of one measure of the best fresh-burned Merstham, Dorking, or other equal approved lime, one measure of finely-ground genuine Italian pozzuolana and two measures of sharp above-bridge river sand, all clean and free from rubbish, dirt and other impurities.'

Two months later the same contractors received another contract—'for a Further Portion of the Foundations and the Sewer'. In this the concrete was to a different specification—six measures of gravel and sand to one of ground stone lime; and also the mortar—three measures of sharp river sand to one of lime, 'the whole well mixed in a pug mill'. All the brickwork was to be laid in English bond.

A month after Barry's wife laid the foundation stone in April 1840, the Government gave the third contract. This was 'for completing the carcase of the River Front, and the North and South Flanks of the building'. On Barry's recommendation eight builders were invited to tender. The highest tender was £184,639. The lowest from Thomas Grissell and Samuel Morton Peto was £159,718 and they were given the contract. Cubitt also tendered but his figure was too high. In this contract Barry had the right to remove Grissell & Peto's foremen, gatekeepers, or watchmen if he thought them incompetent or improper in any way. He did not exercise this right during the strike of the masons in the following year, described on page 57-63. The walls of the Houses of Parliament were to be started at 2 feet 3 inches above high water level as they were brought up to that level.

Grissell & Peto also received the next four contracts, again on Barry's recommendation. The fourth contract of December 1840 was for the 'Foundations for the New Houses of Lords and Commons, the Central Hall, The Royal Staircase, etc.' The estimate was £16,530 and Barry was requested to give reasons for recommending the same contractor. His arguments were sound enough; that Grissell & Peto had already shown their efficiency, their previous competitive tender had been the lowest, their prices were reasonable anyway, and that it was better to avoid having two different sets of workers under different employers who would get in each other's way. He gave examples of their prices:

Digging and disposing of material	2s. 11d. per cubic yard
Concrete or gravel and lime	4s. 9d. ,, ,, ,,
Brickwork in mortar	£11. 5s. per rod

The Government agreed on condition that the work was done on the same terms as those of the Lee Brothers for the first contract.

The next contract, the fifth, in July 1843, was a much bigger one. It was for 'The Superstructure upon the foundations comprised in contract No. 4', of which the estimated cost was £212,249. Two years later 'the site of St. Stephen's Hall, etc., forming the public approach from the south end of Westminster Hall to the Great Central Hall of the New Palace of Westminster', was nearly cleared of all obstruction sufficiently for a contract to be made for erecting the carcase of this portion of the 'New Houses of Parliament'. This contract, number six, was for £39,631. And the same year, the seventh contract with Grissell & Peto followed. But this was different from the others. It was for 'the internal finishings of the Building Generally', and the estimate was also on quite a different basis. The contractors had to agree to prices for the different kinds of work, which would be at a level *below* the official prices in the Department of Woods and Works which it had obtained by competition for other work. Barry thought they could do this because of the saving which could result from using a new machine for making mouldings and wood carvings. He negotiated the percentage reductions in the Official Schedule of Prices as follows:

	Per cent
Carpenter's and Joiner's work	22
Bricklayer's works	15
Mason,s ,,	18
Plumber's ,,	14
Slater's ,,	17
Plasterer's ,,	12
Painter's ,,	10
Glazier's ,,	20
Smith's ,,	22
Ironmonger's ,,	33½
Pavior's & Labourer's work	12

However, Grissell & Peto had been pressed too hard. Six months later they said they wanted to terminate the contract as 'the remuneration was unsatisfactory'. After a thorough investigation Barry agreed that they were working at a loss. He, therefore, proposed new prices which he recommended as reasonable because they gave a profit of only 8% to 9% on prime cost. The contractors agreed and the Treasury approved.

These contracts specified in the greatest detail how the work was to be done and how it was to be charged. Here are a few extracts:

'The bricks are to be laid with close joint, properly bonded, and well flushed up, solid with mortar, and the filling in bricks are to be subbed in and drawn together, so as to make each joint flush and full of mortar, or otherwise to be grouted as directed.

The flues to the whole of the fireplaces are to be 14×12; those for the kitchen chimneys 14×14; and the whole are to be well pargetted.

594 rods, 194 feet of reduced brickwork laid in cement in lieu of mortar, to be provided to be used in the fire-proof or other arches, and in such other parts as may be desired.

The whole of the walls of the building are to carried up together, and from time to time to finish level throughout, at least every 6 feet in height.

The thin walls, or such as are not carried up to the full height of the building to be worked at their junction with the main or thicker walls into a chase or indent, if directed.

The whole of the walls to be effectually covered from time to time with boards or otherwise, wherever the Architect shall direct.

The fire-proof arches are not to be built until the building is covered in; and prepared corbel courses properly cut to the required radius must be left to receive them.

CARPENTER.

The hips, ridges and vallies are to be of 2-inch deal, 12 inches wide, and rounded where required.

The roofs are to be framed in all respects according to the form, scantlings and method shown or described in the drawings.

The gutters are to be of $1\frac{1}{4}$ inch deal, well secured to proper bearers and upright, the bottom to be laid with a 2-inch current and with 2-inch dips, as shown in the drawings.

Proper dovetailed cesspools to the entire width of the gutters, made and fixed in the several situations shown in the drawings, and are to have a hole cut through the socket pipe.

Provide 32 ornaments to dormers; the value, including fixing, to be twenty-five shillings each, prime cost.

The floors, partitions, ceilings, joists, binders, &c. are to be framed in all respects according to the drawings, and as will be directed.

Allow for thoroughly saturating the whole of the fir that may be required, after having been sawn into scantlings, with Kyan's patent solution for the prevention of dry-rot, &c.

All proper centering to arches, apertures, &c. that may be required are to be provided and securely fixed, and are not to be removed until directed by the Architect.

MASON.

The plinth of the external walls, as shown by the drawings, is to be of the best granite from the Fogging Tor Quarries, Devonshire, from Penryn, or from the Island of Guernsey, of a fine grain, similar to that of the best Aberdeen, or from Aberdeen, and of an even colour throughout, being entirely free from large crystals of quartz or felspar, as well as from redness or stains.

Every stone is to be fine axed on the external face so as to present a fair and perfectly even surface, at least equal to a sample now upon the ground, which has been approved by the Architect.

The beds and joints are to be full and square for their whole depths, particular care being taken to preserve the outer arrises, so that when the work is set, it may be close and solid throughout, without any packing; and no joint is to exceed one-eighth of an inch in thickness.

The backs are to be picked square with the beds, so as to present a fair surface against the brick backing.

The granite work throughout is to be laid on a thin bed of mortar prepared as hereafter described, the face of the joints being pointed with cement, and the rest grouted full with mortar.

All the stones are to be worked on the ground and set with lewises and proper tackling; all the arrises and angles of the buttresses are to be very carefully kept, and the lines of the moldings, the curves and the mitres (which latter are in all cases to be worked out of the solid) are to be chiselled to their true and proper form.

The plinth of the elevation towards the courts at the back of the river front is to be likewise of granite of the best quality, worked and set as before described; the dimensions and form are to be as shown on the drawings, the beds to be $16\frac{1}{2}$ inches in width, and the length of stones may vary from 3 feet 6 inches to 7 feet.

The mortar for the granite work is to be composed of one measure of fine ground genuine Italian pozzolana, two equal measures of clear river sand, and one equal measure of good fresh burnt Dorking lime mixed with water and well worked to a proper consistency, and in that state ground with edge stones.

The external walls of the building and all such parts thereof as may be seen from any point of view externally are to be faced with either Portland, Darley Dale, Bolsover, Anston or Steetley stone of the very best and most durable quality, and of an uniform colour, free from all loose vents, holes and all other defects, or with stone from the immediate vicinity of the above-named places, of similar quality and colour.

The stone for the facing and decorations of the walls of the internal elevations towards the courts will be either Mansfield, Haydor or Steetley, and for all the decorative masonry in the interior of the building will be either Haydor or Steetley.

The external face of the work throughout, including the dressings, is to be fine-chiselled. equal to a sample approved by the Architect now upon the site of the building, or rubbed, and the whole of the face of the internal work to be worked fair and rubbed or fine-chiselled.

The plain work in all cases to be true and out of winding, and the mouldings carefully worked to preserve their true form and arrises.

The beds and joints to be random-chiselled, fair and out of winding, and to be sunk wherever necessary to make good work.

The backs of the ashlar and dressings to be roughly drafted square with the beds.

Every stone to hold its full length and height square to the back, and to be secured to the brickwork of the backing by iron cramps at right angles with the face made out of saw plate 1½ inch wide, well pitched and sanded, and of such lengths as to turn down into the stone 1 inch within 4 inches at least from the surface, and up into the brickwork 2 inches, to the extent of 9 inches from the back of each stone respectively; the mortices in the stones for the receptions of such cramps to be filled up and flushed with Roman cement.

The verticals joint throughout to have a rough joggle 10 inches deep from the top bed, and 1 inch square, thus, $\overline{\diamond}$ in which is to be dropped one or more stone pebbles about ⅞ diameter, when the whole cavity is to be filled and flushed up with liquid Roman cement.

Every attention is to be paid to the bonding of the work in the most perfect manner, according to the drawings.

The pinnacles and turret tops are to be in single stones in each course, and every bed is to be plugged together with at least 2-inch square dovetailed iron plugs, 10 inches long, faced with cement.

The horizontal and arch beds of the mullions and tracery of the windows to be also secured each with a hard stone plug 1½ inch square and 2 inches long, carefully faced with cement.

Provide and fix all necessary arch springers that may be shown on the drawings.

All the chimney openings are to have 3-inch tooled Yorkshire landings, 4 inches wider than the depth, 18 inches longer than the length of the openings, with hole sunk through the same for the flue, except the kitchen and other equally large openings, which are to be 4 inches thick.

The throats of the chimnies are to be gathered over from the jambs in the usual manner, and half-brick walls are to be built on these landings to form the flue from the top thereof until it joins the bottom of the chimney throat.

The carving will be done under the exclusive control and management of the Architect and the cost is to be included in this contract; but the Contractor will have to square the stone in the same manner as if the carving were part of his contract, and he must allow the use of his scaffold, &c. for the use of the carver when required.

The Contractor must provide for the security of the masonry from all damage; and all injury that it may sustain during the construction of the building, from whatever cause it may arise, must be repaired to the satisfaction of the Architect, so that it may be delivered up in a complete state when the building is finished.

The metal frames which are intended to be inserted in the stone mullions of the windows are not to form any part of the present contract, except that the grooves for the same are to be formed.

The Contractor must also include the expense of cutting holes in the stone for fixing all iron work, and any other purposes required for the execution of the works generally.

All projecting cornices, strings, &c. are to have water joints run with lead.

All proper plugs, cramps, &c. to be used to the extent hereinbefore provided.

Cut grooves for lead flashings in stone parapets.

PLUMBER.

All the gutters are to be formed as shown by the drawings, &c., lined with cast lead weighing 8 lbs. to the foot superficial, laid to a current of 2 inches to 10 feet, with drips 2 inches high. The lead to turn up to the top of the gutters on each side.

Put lead flashing, 6 lbs. to the foot, round the parapets, turrets, pinnacle, shafts, gables, &c. 6 inches wide; those to the gables 16 inches wide, securely wedged into the stone-work with lead wedges, and pointed with cement.

The flats are to be covered with 8 lbs. lead, to be carefully dressed round the wood rolls, 2 feet 6 inches apart, with all proper laps, nails, screws, &c. complete, and turn up 6 inches against the parapets.

Put 6 lbs. lead flashing round the same where required.

The lead to the dormers to be 7 lbs.

All the lead must be cast of the best Newcastle lead, marked WB.

The flashings to be milled.

It will be required that all the lead shall be cast from the pig in the presence of the Architect, or some person appointed by him for that purpose.

SLATER.

The roofs are to be covered with the best Westmorland or Cumberland slates, of a light greyish green colour, secured with three copper nails to each slate, cut close to all ridges, corners, chimney, shafty and gables; the slates to be so laid that every third slate shall up over every first slate at least three inches, and the eaves to be laid double.

SMITH AND FOUNDER.

All the cast-iron work must be of perfectly sound and smooth castings, correctly placed, bedded and run with lead at their respective proper levels.

All the castings are to be made at some foundry in London approved by the Architect, and are to be placed in the stove-room and heated to a high temperature, in which state they are to be covered with linseed oil, which is to be well rubbed in.

All the wrought-iron work is to be of the best English iron, and the quality and strength of every article, particularly of the castings, must be submitted in the presence of the Architect, or some person appointed by him, to a full proof, at the expense of the Contractor, before it is removed from the foundry.

Provide 20 tons of cast-iron in girders, &c., beyond what is shown on the drawings, including the requisite molds to be used where directed, including 300 yards superficial of white paint.

Provide and fix an ornamental cast-iron trellis grating, each foot superficial weighing oo lbs. to be dropped into rebates over all the gutters, and to be cast in lengths of 3 feet, slaked in linseed oil as before described, and paint the same three times over with the best white lead and oil colour.'

The same contractors, Grissell & Peto, constructed most of the great building between 1840 and 1852. That year, when the Commons occupied their chamber, Queen Victoria formally opened the new Houses of Parliament and conferred a knighthood on Barry. From then on the contractor was John Jay. His work was mainly to finish the two tall towers, Victoria Tower and the Clock Tower. Victoria Tower had reached a height of 160 feet when he started and was to reach 325 feet six years later. In February of that year (1858), 'the Smiths are hard at work hammering together the iron roof which is to form the termination'. However, Jay had found the contract unsatisfactory and was quite willing to give it up if he was compensated; and he refused to tender for the rest of the works.

The labour force was very large, though it varied over the years between 400 and 1400. For instance, in February 1848 there were 1,399 men employed of whom 776 were on the site, 120 in quarries, 335 at the workshops, and the remainder on miscellaneous works. Also, the continually increasing cost of the building was always being criticised. Thus there were attempts to reduce the time required for operations and the labour needed. Barry introduced some of these improved methods.

First a word about materials. They were mainly traditional ones, stone, brick, timber. This, and the enormous amount of detailed decoration, outside and inside, limited improvements in methods of work. The building was, however, a landmark in the use of cast iron, a vast amount was used. Cast iron had been used in construction for some years, but it was only forty years since the first structural cast iron beams had been used in a flax mill at Shrewsbury. Since then the need to reduce the risk of fire had resulted increasingly in cast iron beams and columns replacing timber. Many engineers had to experiment to get the right shapes and sizes. At the Houses of Parliament cast iron was used on a greater scale than ever before. Fire proofing required iron girders and columns carrying brick arches. The roof, including the covering plates, and the framework of the great clock that strikes Big Ben, were also of cast iron. Even the flagstaff on the Victoria Tower, was of rolled sheet iron, 110 feet long and 3 feet in diameter at its base.

One method of reducing cost has already been mentioned, the 'newly invented machine for preparing mouldings and wood carvings'. The Government spent £20,000 on workshops at Millbank for this and other machines. At one time there were over 300 men working there, though only a few could have been on this machine. Another time-saver was 'Doctor Spurgin's Machine for Hoisting Bricks, Mortar, etc.' Here is a drawing of it. The letters I, K and L show different containers for broken brick, water and pieces of stone. The Doctor claimed that by making it unnecessary for workmen to mount ladders accidents were avoided and operations were quicker and cheaper. Cubitt used the machine as well as Grissell & Peto.

But the most important innovation was 'the Mechanical Scaffolding'; a system of 'so-called whole timber or framed scaffolding, with its tram-way and crab engines aloft'. This was not in fact new but it was still unusual and had only begun to take the place of the old scaffolding of poles and ropes. As with many other things, its wide use at Westminster, encouraged the industry to use it more. First used on the river front the system was adapted throughout the whole building and was particularly useful on the three tall towers.

The basic problem of handling was to raise to considerable heights a great number of blocks of masonry each weighing four to five tons. These were elaborately carved as a result of three months' work in the workshops. They had to be handled with care and placed at some lofty point where they could be fitted into position like a brick. It was to meet this problem that a system of tramways under and on the scaffolding with travelling trucks and hoisting steam-engines was introduced.

On the first tower to be completed, the central tower, 266 feet to the top of the spire, 25,000 cubic feet of worked masonry was moved in this way. The clock tower, 211 feet to the top of the stonework plus another 103 feet to the iron roof of the bell chamber, was a bigger job. It measured an average 40 feet square. Thirty thousand cubic feet of stone, 300 rods of brickwork, the iron frame for the clock, and the beams and plates for the roof were lifted in this way. The $2\frac{1}{2}$ h.p. steam-engine on a platform at the top raised the material in a shaft inside the tower, later used for stairs and lift, so that no scaffolding was to be seen outside. The platform which with its engine, crab, trucks and tramways weighed 16 tons, could easily be raised to the height required. The same method was adapted for the even bigger Victoria Tower. In this case the movable platform with its load weighed some 40 tons and greater strength was necessary. There was:

'a strong trussed frame, constructed to carry all the machinery, consisting of balks of timber, 51 feet long and 14 inches square; these beams crossed the area and similar pieces tied all round close to the inside walls; diagonal braces at each corner tied together; and four beams crossing the centre of the tower strongly trussed, above and below; the whole perfectly rigid and stiff as proved by constant use for nine years. A circular cast iron rail was next laid on the framed platform, while part of the framing was covered with $2\frac{1}{2}$ inch planking, and defended by a hand rail for the safety of workmen, and on this the portable engine of six-horse power was placed, with its drum connected by gear work with the driving wheel of the engine.'

On the Victoria Tower it was not possible to raise the material inside the tower above the height of 63 feet as there was a stone ceiling at that point. It was therefore necessary to bring it up outside the walls by means of a framing or traveller, parallel to the platform and extending over and clear of the walls. This method was used up to a height of 250 feet, 6 feet above the parapet. Above this rose the pinnacles, another 85 feet. For these, cradling scaffold was needed. This was considered to be 'very daring in its construction since it was entirely detached from and independent of the masons' work and in no way resting on or touching the work'.

The 'mechanical scaffolding' made great savings in time and labour, and it was safe. On the Victoria Tower it raised 117,000 cubic feet of

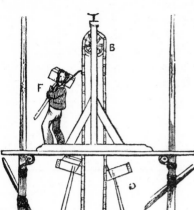

To explain the immense advantage which the machine offers in expediting the work of building, and diminishing the expense of raising the bricks and mortar, it seems to be unnecessary to do more than to refer to the following.

DYNAMICAL TABLE OF THE STRENGTH OF A MAN

Height.	A Minute.	An Hour.	Ten Hours.
to 10 feet	90 bricks	5,400	54,000
,, 20 ,,	45 ,,	2,700	27,000
,, 30 ,,	30 ,,	1,800	18,000
,, 40 ,,	22 ,,	1,350	13,500
,, 50 ,,	18 ,,	1,080	10,800
,, 60 ,,	15 ,,	900	9,000

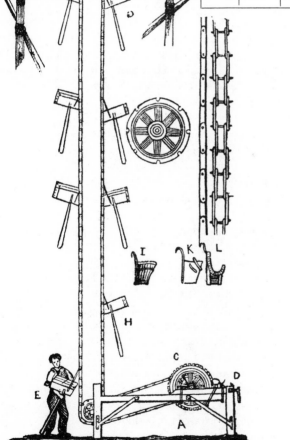

OR SPURGIN'S MACHINE FOR HOISTING BRICKS, MORTAR, &c.

stone, 1,350 rods of brickwork and 190 tons of iron. It often had loads
of 40 tons. The average time of transit from ground to 250 feet,
including deposit on the platform, was 3½ minutes; the average number
of lifts per hour was 10. The cost of the steam engine with all the tackle,
which used three hundredweight of coal in its normal working period
of five hours, was £1,800. But the mason setting stones high up on
the tower could be kept continually at work. Barry himself provided
the scaffolding on the towers since the contractor had refused to be
responsible for it.

A building so much in the public eye and so important for the
Government of Britain was bound to arouse a great deal of comment
and criticism during the long years of construction. Some people
disliked the Gothic style. But most of the criticism by both Parliament
and the public was about the cost and the delay in completion. These
two could hardly be separated. For instance, the problem of the
foundations increased the cost and delayed the start on the structure.
For the lofty massive bulk of the Victoria Tower, of which the site
was full of quicksands—

'A double close row of piles was driven, completely surrounding the
site, the earth within this species of coffer dam was then excavated and
concrete thrown in to a depth of 10 feet 7 inches. Two courses of six
inch landings were then placed on the concrete and upon them the
walls were carried up, without accident of any kind'.

The original estimate of cost in 1837 was £707,104 but by 1858,
when the building was nearly finished, £1,768,979 8s. 1d. had been
spent. These figures needed some explaining, but there were a number
of good reasons for the enormous increase. One factor outside anyone's
control was the slight but continual rise in prices and in wages during
the eighteen-fifties. But the real reason was that the building as it was
when finished was very different from the one specified in the original
estimate. Furniture and fittings accounted for half a million pounds of
the difference between the two figures. Neither heating and ventilating,
nor construction of the embankment, nor fire proofing, were included
in the estimate. Another unforeseen factor was that the Anston stone
which had to be used was much harder than that selected at first.
There were also additional parts of the building, the central tower, and
several residences for officials. Barry himself was responsible for some
of the extra cost. He was too optimistic about the difficulties in
carrying out his plans and, in his desire to have only the very best, he

repeatedly altered the designs of different parts, for example, the clock tower.

The time taken in construction seemed quite unreasonable to contemporaries. Complaints about the delays appeared continuously in the press. The papers published verses about the subject, in which Barry rhymed with tarry. On the stage too, in a show at the Adelphi theatre a character had certain plans which he was:

'Long in getting through with
Like certain Houses Barry has to do with'.

Such attacks went on all over the country. Only four years after the first stone was laid the House of Lords became concerned about the slow progress and appointed a special committee of enquiry. The Lords occupied their chamber in 1847. Then it was the turn of the Commons. Repeatedly Members rose to complain about the delay and the cost, and in 1848 the Government felt obliged to appoint a special commission to 'hasten completion and superintend expenditure'. The subject became a joke. Two years later a question in the Commons caused laughter all round. There were endless debates about the internal decoration of the chamber. In fact Parliament itself was responsible for much of the delay. It was mean about expenditure on public buildings. It held up grants of money for the works, especially as the cost rose. In 1850 the Government owed the contractors many thousands of pounds with the result that many men had been sacked, the site was practically at a standstill; and the workshops were empty except for £14,000 worth of materials lying around. Two years earlier the labour force had actually been reduced from about 1,000 to 500. At last in 1852 the Commons moved into its new chamber after numerous alterations had been made: the ceiling lowered five feet to improve the acoustics, the public gallery enlarged, and the Speaker provided with his private hot water plate to warm his feet. The Victoria Tower was still only half built and there were as yet neither clock face nor Big Ben.

Barry's troubles were by no means over. M.P.s complained that they got headaches in their new chamber and suffered from draughts from all directions. The problem of ventilation and heating of such a great building dragged on for years. The Government had appointed a so-called expert, who was not responsible to Barry, to deal with it. But he was a medical man, not an engineer or architect, and continually clashed with Barry with the result that much money was wasted. The Government set up another select committee which merely reported that there was not enough knowledge on the subject. The big clock and Big Ben were a long tale of woe. The job of designing the clock was passed by one expert to another because of its unprecedented size, and when it was finally fixed in 1859 the weight of the hands stopped it. The great bell was more unlucky. The first Big Ben, cast in 1856, cracked under the impact of its clapper before being hoisted. The second also cracked after being hoisted three years later. Along with Big Ben it was also necessary to tune the quarter hour bells as follows:

	Diameter at mouth	Weight cwt.	q.	lb.	
1st quarter	45″	21	0	0	A flat
2nd ,,	48″	26	0	0	G ,,
3rd ,,	54″	32	0	0	F ,,
4th ,,	72″	77	3	24	C ,,

Before this there was already a problem of decay of the stone which had begun about seven years after building started. £7,000 was needed for preservation and the great scientist Michael Faraday gave his advice. No one would take the blame; the Government experts had certified the quality of the stone but it was alleged that the contractor had taken the stone from the Duke of Leeds's quarry where it was inferior and unsound.

The Palace of Westminster was not completed until five years after Barry's death in 1860. His son had made the drawings for the top of the Victoria Tower. Barry himself had had to carry through the vast project, in continual conflict with Government authorities and experts of different kinds about money, estimates, contracts, internal decoration, heating and ventilating and about his own fee which had been fixed at £15,000 instead of the more usual 5%.

At last in 1865 the Houses of Parliament with their 1,100 rooms, 19 halls, 126 staircases and 2 miles of corridors and passages was finished. Two years later Parliament decided to give the vote to working men in the towns, the men who had built its Houses.

THE BUILDING WORKERS (1835–1860)

After the collapse of the great Operative Builders' Union in 1834 the building craftsmen led a quieter life for many years. There was, of course, a big expansion of building operations during the following twenty-five years. The economy grew more rapidly than ever before and so a great number of railway stations, factories, and town halls were built. Consequently the number of craftsmen increased. In 1851 there were 443,000 of them in a population of twenty-one million. By 1871 this number had risen to 630,000. Their work was in great demand, except during the slumps, but they did not take advantage of this situation for a long time.

There were several reasons for this. When the O.B.U. disappeared the effect on the national unions of the separate crafts was disastrous. Those of the plasterers, painters and slaters disappeared. Those of the stonemasons, bricklayers, carpenters, and plumbers just survived, but very much weaker. Another reason was that during the eighteen-thirties and 'forties the most active members of the working class gave their time and energy to the Chartist movement. Although the stonemasons' union would not support the Chartists officially it did recommend their newspaper *The Charter*, to its members. The carpenters were more active in that movement which lasted from 1836 to 1848.

The Operative Stonemasons' Society was the only strong and stable union during all that time. In fact it was the most powerful union in the country until the rise of the engineering workers. There was still a great amount of building in stone, as in the Houses of Parliament; the stonemasons were the only body which could act nationally; the other crafts did act, but in each town separately, in London or Birmingham. But for all of them the watchword at first was caution. The stonemasons' central committee said:

'When will our caution against strikes have the desired effect? Will our members be convinced of their destructive tendency, and abandon the thought, or will they continue to cut off every sprout of prosperity as it makes its appearance amongst us?'

At the same time the stonemasons set about strengthening their organisation. The black list was started; a list of members who had acted against the union in some way, which was circulated so that they might be stopped from working. The administration and finance (the subscription was sixpence a week) were centralised at head office. By 1838 the membership had recovered and risen to 4,953. It was just as well, for three years later the union entered the biggest struggle it ever had to face alone.

On Monday, 13th September 1841, 468 stonemasons working on the new Houses of Parliament went on strike against the foreman,

George Allen, 'a man who damns, blasts and curses at every turn'. Two months earlier the contractors, Grissell & Peto, had received a deputation and thought they had settled the grievances. They agreed that the supply of beer which they had restricted to the two public houses nearest the works should come from the house where the masons usually held their meetings. Another grievance was that Allen had refused a man leave to bury his mother, but enquiry showed that the man wanted three weeks off. But on a third and most important point they had refused to give way: the amount of a fair day's work. The union had laid down the quantity of work to be done by each man in a given time. One mason had finished the working of a window head in two days less than the time stated. Such men were called 'chasers' by their work mates. The masons wrote to the culprit as follows:

'From the Masons to William Geggie, one of the masons employed at the New Houses of Parliament.

Brother Geggie, I am instructed by a meeting of the masons employed on this job, held at the Paviors' Arms this evening, to which you were summoned, but failed to attend, to inform you that it was unanimously resolved that you should pay the fine of 5s. for your late conduct in chasing. You must pay the fine to the mess steward (Thomas Downs) by Saturday night, August 4, 1841, at the latest.

Signed on behalf of the meeting

John Worthington, Chairman.'

Bro. Geggie refused to pay when supported by Allen acting on the contractors' instructions. The masons therefore accused Allen of trying to get more than a fair day's work from them, and of using intimidation. On the other side the Government department which had given the contract (the Commissioners of Woods and Forests) promised the contractors its support. On Tuesday, 21st September, at a meeting of 200 masons at the Craven Head, Drury Lane, it was resolved to continue the strike and get the support of other masons in London. A further complaint now was that Allen had locked the pump in the Palace Yard, thus forcing teetotallers to go to the public houses. Three days later the contractors showed their hand. In a notice placed on the hoarding round the site they said that the men were led astray by their leaders, that they could not dismiss Allen, and that they 'are collecting masons in all parts of the Kingdom' and therefore the men would find their jobs filled unless they returned quickly. Next a Member of Parliament for Westminster, Captain H. S. Rous, tried his hand at mediating. He saw both sides; the masons' case was still that Allen had 'committed acts of tyranny and oppression' and they were determined not to work under him; the contractors admitted that Allen was 'a rough diamond, not particular in his expressions, but a man who has received a university education was not adapted for a foreman of masons' and they also refused to budge. That attempt having failed, a day or two later a well-known builder, named Jackson, offered to sort the matter out with the men at a meeting at the Paviors' Arms, Millbank, their headquarters. But he also failed. He tried to divide the meeting by asking whether the leaders had not misled the men by alleging that they would undergo more severe discipline if they returned; and he left the meeting in confusion, no doubt as he intended. But the next evening the meeting stood firm. Members claimed that if only they stood together they could ruin the largest builder. They recalled proudly the Dorchester labourers whom the Government had been forced to return from Botany Bay.

Both sides braced themselves for a struggle. Messrs. Grissell & Peto advertised for masons in Newcastle but they received a stern reply:

'To the Master Builders of the New Houses of Parliament, London. Gentlemen (miscalled),

The masons of Newcastle-upon-Tyne beg to inform you that they have very reluctantly read your falsified bills, posted in all parts of the town, purporting to be in want of hard stonemasons to work at the new Houses of Parliament, while indeed they are not wanted! All that is wanted is an equitable adjustment between employer and employee; anything more is gross falsehood, also a paltry allurement to drag men from their peaceful homes to be the stupid tools of a rascally faction; hence the destruction of our brother operatives in London. We would further inform you that your weak efforts at delusion have no power whatever here; and we would recommend you to agree with your own workmen, as we do assure you that instead of coming to take the employment from our fellow workmen, we have determined to give our last shilling to their support. The labourer is worthy of his hire and the mechanic such wage as becometh his scientific operations. Hoping you will not again disgrace the walls of Newcastle with your delusive, transparent, paltry placards, we remain with all due respect to our brother workers,

THE STONEMASONS' SOCIETY OF NEWCASTLE UPON TYNE'

In response to this publicity and the public interest in the Westminster site on 2nd October the contractors sent their case to *The Times*:

'Charges made by the Men.

'Charge 1.—One instance is of a man who unfortunately fell from the scaffold and broke his leg, who, as soon as he was able to resume his employment, returned to his work, expecting at the least to find a compassionate feeling would be shown towards him; but, however. no; he was brutally told to go about his business, as he (Mr. Allen) did not want such men on the work.

'Charge 2.—Another instance is that of a man who was taken ill in consequence of his being previously out of employment, and unable to obtain the common necessaries of life, and having at that time a wife who had been ill for some months; he was in consequence obliged to remain from his work a few days, and when he came back, expecting to resume his employment, he was insultingly told "to leave the ground," as he (Mr. Allen) wanted none there but sound men.

'Charge 3.—Another instance of his cruelty and tyranny was exercised upon an individual, who, obtaining leave from an under-foreman to remain at home to close the eyes of his dying wife, was told the ensuing morning when he came to his work by this Allen "to go and die with her and be d——d."

'Charge 4.—Another is a young man who received the distressing news of the death of his mother, and consequently solicited Mr. Allen to allow him to go and see her buried (the time required was a week or a fortnight), but this unnatural being unfeelingly told him he might go, but he should not come back again, as he would not keep his work open for any one under such circumstances.

Grissell and Peto's Refutation.

It is perfectly true, that a labourer (not a mason) had the misfortune some twelve months since to break his leg by the falling of a stone upon him, and upon his recovery he applied to our firm for relief, and a recommendation to enable him to obtain some other less laborious employment. The statement that he applied to us again for employment and was refused is not true.

'As to this charge, neither Allen nor any of the under-foreman have any knowledge or recollection of the circumstance.

'In respect to this, all our foremen agree in declaring that the circumstance related never could have occurred. No man would have been discharged under such circumstances; the enormity of the charge carries with it its own contradiction.

'With respect to this, we can bear our own personal testimony to the untruth of the statement. The circumstance was simply this: a man applied to the foreman for leave of absence to bury his mother, to which the foreman immediately assented, but upon the workman mentioning that he should be absent about a fortnight or three weeks, the foreman replied that he could not undertake to keep the place open for him so long a time; the man then replied, if that was the case he would not go at all. We believe he did go, and returned to his work, at which he remained until the time of the strike.

They also stated the real point at issue:

'The real object of the few who govern the many is to show to the public, and the builders in particular, the power which the combined union possesses to control the price of labour and regulate the quantity of work to be performed, to an extent which can only be anticipated with alarm by the builders in the first instance, and which must ultimately operate upon the public at large, by necessarily raising the prices of every article connected with building; such a power must likewise effectually put an end to public competition, for what builder of capital would jeopardise his property in large public contracts, calculating as they must do upon the ordinary value of labour, if workmen during the progress of a large undertaking are to become the judges and dictators to their employers of the quantity of work that is to be done, and to have the nomination of their own superintendent?

'Another serious influence of the union is, that every workman, good or bad, idle or industrious, shall receive the same rate of wages, namely, 5s. per *diem*, and that no man is permitted to undertake task-work, by which absurd and unjust regulation hundreds of workmen are prevented from availing themselves of their own capacity and industry to provide for their families and against the contingencies of sickness and old age.'

A week later the result of the contractors' efforts was seen when 16 blacklegs arrived on the site from Cirencester, and were greeted with loud mocking cheers from the masons, and after another week a further 60 arrived. These only did rough work and many of the stone carvers also struck against working with them. In the meantime the union had also been busy on Grissell & Peto's other jobs. These were the construction of Nelson's Column in Trafalgar Square and the new steam dock at Woolwich. The men there said they would strike unless the demands of their fellow workers at the Houses of Parliament were met. In spite of a meeting in Trafalgar Square with Grissell himself they carried out their threat. But the Admiralty told the contractors that they would not press them to complete the works.

By the end of October many more blacklegs had been recruited on all three sites. 'Many excellent granite masons' were expected from Scotland. All the same the Scottish Stonemasons' Society sent the strikers £500. At Woolwich Grissell & Peto announced that they would re-employ the masons on strike if they applied quickly.

However, the strike continued throughout the winter. This was largely due to the masons' energetic secretary, Thomas Shortt who was also a Chartist. He issued a regular bulletin and large posters, and organised mass meetings which were addressed by Tom Duncombe and Wakley, the radical M.P.s for Finsbury, and by Feargus O'Connor, the great Chartist orator. He organised committees all over the country, collected nearly £5,000 for the strikers and thoroughly aroused public opinion in London. The agitation was carried on jointly by the masons and the Chartists. Earlier that year eighteen stonemasons had carried into the House of Commons the petition for the release of imprisoned Chartists. In any case there was great interest in Nelson's Column where work stopped completely for several weeks and cartoons in the papers made fun of the situation.

All the same it was the determination of the masons which kept the strike going so long. They felt it was as much a question of liberty and of their dignity as of wages. This feeling was expressed at a meeting on 5th November in the Crown and Anchor. Their motives were misunderstood said various speakers; they struck not for an increase in wages but against the 'tyranny exercised by Allen'. The newspapers were condemned for their prejudice and for not making clear that it was a question of human rights and freedom, not of wages. 'This was the most intellectual strike that had ever taken place in the trade.' A carpenter and a Chartist urged the meeting to support only the Chartist paper. Another speaker claimed that Allen himself had been sacked by Cubitt's in 1834 because he was a union member but now he was 'acting the part of a petty tyrant'. Then the M.P. from Finsbury roused the meeting:

'The workmen must not expect any support from the aristocracy, for whom their labour created the capital which now crushed them. On the spot where the masons struck was to be found the source of the evils they complained of. Their object should be to produce a reform of the legislature . . . the ground upon which they now stood out was the highest ever yet taken by the working classes. It was not a question of wages, but whether they should be treated as beasts of burden, or as moral and intellectual beings.'

But by the beginning of 1842 the union had given up hope of winning. In February at the Dover Castle, in Commercial Road, 'a number of tradesmen' among whom were several very 'wealthy men' presented a silver snuff box, value £20 to George Allen, the foreman under dispute. One of them, an 'extensive lighterman', no doubt concerned with the transport of materials by river to the site, made the presentation as a 'mark of respect because of the integrity and firmness shown during the strike because of his refusal to be subjected to the tyranny and injustice of the union. . . .' In April Prince Albert was taken by the architect on a tour of the site. By May 1842 all the strikers were re-employed and the strike was declared off. In that year there was a heavy slump in trade; in Bolton two-thirds of the masons were unemployed.

In spite of this defeat the stonemasons were not crushed. Only four years later they won a victory in Lancashire, but this time jointly with all the other craftsmen in the industry. They debated whether to demand higher wages or shorter hours. One mason argued:

'By reducing the hours of labour in the summer we shall secure plenty of employment during the winter. It has been stated that the lives of masons are shorter than the men in other trades. I am aware that this is the case. I have seen young men go off very quickly. What is the cause? The cause is hard labour. . . . When a mason comes to about forty years of age he is generally troubled with a cough—he goes to a medical man and states his case—the doctor shakes his head and says: Well my man, I have had several cases of this sort. It is the masons' disease; all I can do for you is to give you some temporary relief— something to ease your breast.'

All the crafts decided to strike for a nine hour day. The painters and joiners soon returned to work but the bricklayers, plumbers, plasterers and labourers stood firm with the masons. The employers locked out every worker in south Lancashire and only after several months made peace. The result was that although the workers did not get the nine hour day, they had defeated the employers' presentation of the Document, and immediately after they did get an extra shilling and half an hour's break for tea. By the end of 1846 the masons' union had risen to 6,000. The other craft unions, however, fell apart into small local societies.

The masons' union owed its continued strength largely to the strong character of their new secretary, Richard Harnott, who was elected in 1847 and filled that post for the next twenty-five years. He increased the power of the central committee and headquarters and reduced the independence of the local lodges. At this time many trade unionists still saw their union as no more than a collection of lodges, the really important units, linked together by its secretary. Against much opposition from the lodges Harnott brought the union nearer to a modern type with a central authority. But the lodges were still important, specially because they provided shelter for the mason tramping in search of work. The system was described by Henry Broadhurst, the stonemasons' M.P. in his book, *Story of My Life*.

'My trades union had relieving stations in nearly every town, generally situated in one of the smaller public houses. Two of the local masons are appointed to act as relieving officer and bed-inspector. The duty of the latter is to see that the beds are kept clean, in good condition, and well aired, and the accommodation is much better than might be expected. When a mason on tramp enters a town he finds his way to the relieving officer and presents his card. On this card is written the applicant's name and last permanent address. In addition, he carries a printed ticket bearing the stamp of the last lodge at which the traveller received relief. He was entitled to receive a relief allowance of one shilling for twenty miles and threepence for every additional ten miles traversed since his last receipt of relief money. Thus, if fifty miles have been covered the man receives one and ninepence. In addition, he is allowed sleeping accommodation for at least one night, and if the town where the station is situated is of considerable size he is entitled to two or three nights' lodging. Besides a good bed, the proprietor of the official quarters is bound to furnish cutlery, crockery and kitchen conveniences for each traveller, so that the relief money can all be spent on food. There is also no temptation to spend the small sum received in intoxicating drink unless its recipient chooses to do so. The system is so perfect that it is a very rare occurrence for an impostor to succeed in cheating the union. Unfortunately, the stations did not exist everywhere, and where they were separated by forty or fifty miles—not a rare occurrence in the southern counties—the traveller's life becomes a hard one. I have frequently had to provide supper, bed and breakfast on less than a shilling, so it may be readily imagined that my resting places were never luxurious hotels.'

Harnott's chief concern was to strengthen the finances of the union.

When money was low he discouraged strikes, when there was enough he said 'We must dispute every inch of the ground with the capitalist and not flinch one iota'. During the 'fifties there were three main problems: piecework, the nine hour day or fifty-four hour week, and persecution by the magistrates. The struggle for the nine hour day is dealt with in Chapter 4. The bricklayers and plumbers as well as the masons had fought against piecework ever since the great days of the Operative Builders' Union. They had rules against speeding up or 'chasing' as at the Houses of Parliament. The bricklayers fined any member for 'running or working beyond a regular speed'. The opposition to piecework also included sub-contracting in which a mason would take on a small job by the piece. They also fought against speeding up under time payment. Harnott wrote in 1850:

'It has come to our notice that tasking [piecework] is creeping in amongst our trade in various parts of England, yea, and in towns where the system has been utterly abolished, for instance Liverpool, Bolton, Warrington and its locality. . . . If there is any chance of succeeding, grovelling employers will try on the tasking system. Perhaps they will at first allow somewhat liberal prices: this will induce some, especially the young ablebodied men, to engage who think they cannot injure their constitution by any means. These will overstretch every sinew to make as much as they can and upon pay day they receive a *few* shillings above the common rate of wages; this will induce more to join them. The system being once established the worthy employer begins to grumble at the high wages they are making, and a reduction of prices will soon follow: then reduction after reduction, until it reaches starvation; an additional draft of ALE or SPIRITS is then required to assist the physical power or in other words to keep the human machine in a state of stupor and insensible of the injuries inflicted upon it by the unwise conductor. But corroding of the lungs, piles, rheumatic pains and other diseases will show themselves, which will surely convince men of their folly, though previously deaf to the persuasions of their best friends to abolish such a horrible system as piecework.

'There is another system these worthies have if they cannot succeed in tasking their work (and we are afraid it is still carried on) which is: they employ one or two of those ignorant animals who are to be found in our order, whose only boast is how much they can work and drink, and who glory how much they can harass their fellow-men; to those found to be so base are allowed two or three shillings extra weekly wages and a bellyful of drink; great respect is paid to them, the best of materials are put into their hands, everything they can say or do is approved of by their employers, purposely to make them work like brutes so as to harass others to keep up with them. The employer or foreman goes about like a *roaring lion* and if he sees any falling behind these BRUSSARS he roars and swears, marking them down to be paid off, or to be paid under the current wages.'

Persecution by magistrates bothered the masons' union considerably. There were many cases of members being convicted on flimsy evidence and sometimes without a proper trial. The offences were usually conspiracy against the masters or alleged intimidation of blacklegs.

The other building unions were much weaker in this period. The bricklayers were split into two Orders, one in London, the other in Manchester. Together this membership totalled only about 6,000, only a tenth of all bricklayers in the country. In London they had a long struggle to win the 4 o'clock Saturday (i.e. to stop work at four on Saturday) and another to raise wages by 6d. to 5s. 6d. a day. The carpenters and joiners were even weaker. There were only a few hundred in their General Union. One employer said, 'If they should ask me for the four o'clock on Saturday, I will make them work till half-past five'.

A London journalist, Henry Mayhew, reported on the life and work of the carpenters in the *Morning Chronicle* during July 1850. He wrote:

'Of the carpenters and joiners now in London, 1,770, or about one tenth of the entire number, are "in society". Their houses of call are almost invariably held in public houses. The objects of these societies are twofold—the upholding of the wages of their trade, and rendering assistance to the aged, disabled, or unemployed or their own body. The members meet periodically at their respective houses of call and contribute such a sum per week (varying from $1\frac{1}{2}$d to 4d.) as is deemed necessary under the circumstances of the trade and the society.'

One of these men told him:

'I have known the London trade between twenty and thirty years. I came up from Lancashire, where I served an apprenticeship. I have worked all that time entirely at carpentering. No doubt I am a pure carpenter, as you call it, never having worked at anything else. . . .

I have always had 5s. a day, and in busy times and long days have made 33s. and 35s. a week, by working overtime. I have always been able to keep my family, my wife and two children, comfortably, and without my wife's having to do anything but the house-work and washing. One of my children is now a nurse-maid in a gentleman's family, and the other is about old enough to go and learn some trade. Certainly, I shan't put him to my own trade, for, though I get on well enough in it, it's different for new hands, for scamping masters get more hold every day. There's very few masters in my line will take apprentices; but I could set him on as the son of a journeyman. If I'd come to London now, instead of when I did, I might have got work quite as readily perhaps—for I didn't get it within a month when I did come; but then I was among friends; but I should have had to work for inferior wages, and scamping spoils a man's craft. He's not much fit for first-rate work after that. I am better off now than ever I was, because I earn the same, and all my expenses, except rent, are lower. I have a trifle in the savings bank. But then, you'll understand, sir, I'm a sort of exception, because I've had regular work, twelve months in the year, for these ten or twelve years, and never less than nine months before that. I know several men who have been forced to scamp it—good hands, too—but driven to it to keep their families. What can a man do? 21s. a week is better than nothing. I am a society man, and always have been.

'I consider mine skilled labour, no doubt of it. To put together, and fit, and adjust, and then fix, the roof of a mansion so that it cannot warp or shrink—for if it does the rain's sure to come in through the slates—must be skilled labour, or I don't know what is. Sometimes we make the roof, or rather the parts of it, in the shop, and cart it to the building to fix. We principally work at the building, however. There's no rule; it all depends upon the weather and convenience. The foremen generally know on what work to put the men so as best to suit, but in no shop I've been in has there been a fixed and regular division of the carpenters into one set as roofers, and another for the other work. Our work is more dangerous than the joiners, as we have to work more on scaffolding, and to mount ladders; but I can't say that accidents are frequent among us. If there's an accident at a building by a fall, it's mostly labourers. I'm satisfied that the carpenters on the best sort of work are as well conducted and as intelligent as any class of mechanics. . . .'

Mayhew contined:

'The journeymen in connection with the "honourable" trade amount, as I before stated, to 1,770, so that by far the greater number, or no less than 18,230 of the working carpenters and joiners in the metropolis belong to what is called the "dishonourable" class—that is to say, nearly 2,000 of the London journeymen are "society men", and object to work for less than the recognized wages of the trade, while upwards of 18,000 are unconnected with any of the trade societies, and the majority of them labour for little more than half the regular rate of pay.'

One of these 'dishonourable' men then told him:

' "I work at what is called a strapping shop," he said, "and have worked at nothing else for these many years past in London. I call 'strapping', doing as much work as a human being or a horse possibly can in a day, and that without any hanging upon the collar, but with the foreman's eyes constantly fixed upon you, from six o'clock in the morning to six o'clock at night. The shop in which I work is for all the world like a prison—the silent system is as strictly carried out there as in a model gaol. If a man was to ask any common question of his neighbour, except it was connected with his trade, he would be discharged there and then. If a journeyman makes the least mistake, he is packed off just the same. A man working at such places is almost always in fear; for the most trifling things he's thrown out of work in an instant. And then the quantity of work that one is forced to get through is positively awful; if he can't do a plenty of it, he don't stop long where I am. No one would think it possible to get so much out of blood and bones. No slaves work like we do. At some of the strapping shops the foreman keeps continually walking about with his eyes on all the men at once. At others the foreman is perched high up, so that he can have the whole of the men under his eye together.

'I suppose since I knew the trade that a man does four times the work that he did formerly. I know a man that's done four pair of sashes in a day, and one is considered to be a good day's labour. What's worse than all, the men are everyone striving one against the other. Each is trying to get through the work quicker than his neighbours. Four or five men are set the same job so that they may be all pitted against one another, and then away they go every one striving his hardest for fear that the others should get finished first. They are all tearing along from the first thing in the morning to the last at night, as hard as they can go, and when the time comes to knock off they are ready to drop. I was hours after I got home last night before I could get a wink of sleep; the soles of my feet were on fire, and my arms ached to that degree that I could hardly lift my hand to my head. Often, too, when we get up of a morning, we are more tired than we went to bed, for we can't sleep many a night; but we mustn't let our employers know it, or else they'd be certain we couldn't do enough for them, and we'd get the sack. So, tired as we may be, we are obliged to look lively somehow or other at the shop of a morning. If we're not beside our bench the very moment the bell's done ringing, our time's docked—they won't give us a single minute out of the hour. If I was working for a fair master, I should do nearly one-third less work than I am now forced to get through, and sometimes a half less; and even to manage that much, I shouldn't be idle a second of my time.

'It's quite a mystery to me how they do contrive to get so much work out of the men. But they are very clever people. They know how to have the most out of a man, better than any one in the world. They are all picked men in the shop—regular 'strappers', and no mistake. The most of them are five foot ten, and fine broad shouldered, strong backed fellows too—if they weren't they would not have them. Bless you, they make no words with the men, they sack them if they're not strong enough to do all they want; and they can pretty soon tell, the very first shaving a man strikes in the shop, what a chap is made of. Some men are done up at such work—quite old men and gray with spectacles on, by the time they are forty. I have seen fine strong men, of six-and-thirty, come in there and be bent double in two or three years. They are most all countrymen at the strapping shops. If they see a great strapping fellow who they think has got some stuff about him that will come out, they will give him a job directly. We are used for all the world like cab or omnibus horses. Directly they've had all the work out of us we are turned off, and I am sure after my day's work is over, my feelings must be very much the same as one of the London cab horses. As for Sunday, it is *literally* a day of rest with us, for the greater part of us lays a bed all day, and even that will hardly take the aches and pains out of our bones and muscles. When I'm done and flung by, of course I must starve.'

FURTHER READING:

E. P. Thomson and E. Yeo, *The Unknown Mayhew*, Merlin Press 1971.

3   Contrast: Cast Iron and Red Brick

This chapter is about the contrast between two designers and their buildings in the mid-Victorian age. The Crystal Palace and St. Pancras Hotel were built within less than twenty years of each other but there could be no greater contrast between them. The difference in materials, methods, style, appearance and in the designers themselves was enormous. The Crystal Palace, built in 1851, was designed by Joseph Paxton; the St. Pancras Hotel, built as part of the railway terminus, between 1868 and 1876, by George Gilbert Scott. The first was of iron and glass, the second of brick and stone.

Of course, there was nothing new about using iron for structures when Paxton designed the Crystal Palace. Sixty years earlier the textile mills had had cast-iron frames to reduce fire risk. A few years later three churches had been built in Liverpool in which the columns, arcades, roof work and the window mullions and tracery were entirely of cast iron. Then in London in the eighteen-forties the Coal Exchange in Lower Thames Street had a large dome which had cast-iron ribs supported by cast-iron columns. A little later wrought-iron became more popular for structures, following the improvement in the production process by Joseph Hall of Timpton. Sir William Fairbairn, the great engineer, argued for its usefulness and the railway companies began to use it for their new termini. At Euston station the roof trusses were of wrought iron, carried by cast-iron beams and columns. This mixed construction of cast iron and wrought iron was, in fact, that of the Crystal Palace. Perhaps Paxton saw the possibility of such a building when he was a director of the Midland Railway.

SIR JOSEPH PAXTON (1803–1865)

Paxton was born at the small village of Milton Bryan near Woburn, where the Duke of Bedford had his stately home. He was the seventh son of a tenant farmer of the Duke. The wall round the Duke's park passed close to the village. In these surroundings and at that time, everyone knew his place and did not think of stepping outside it. Paxton's mother had run away to marry his father, but she had a little money of her own. She seems to have been enterprising and intelligent. Nothing is known about Paxton's education except that he probably learned to read and write at the village school built by the local squire. His father died when Paxton was a boy and he was sent to his eldest brother, also a farmer. The next few years of his life were hard. He ran away from his brother's home into Essex. Later in life he said, 'You never know how much nourishment there is in a turnip until you have to live on it'. Eventually when fifteen he returned to Bedfordshire. There he worked under another brother who was gardener at Battlesden Park, another stately home only a mile or two from his native village.

But from that time he was often on the move. Two years at Battlesden Park were followed by four years at Woodhall Park, near Watton-at-Stone, Hertfordshire, the home of a wealthy city banker. There he was apprenticed to the gardener who was a famous and skilful fruit grower. He returned to Battlesden where he constructed a large lake in the grounds. Wanting to get wider experience he turned to London and for a year worked at Wimbledon in the service of the Duke of Somerset. Then occurred the change from which all his future life developed.

The Horticultural Society had recently opened its gardens adjoining Chiswick House, which it leased from the Duke of Devonshire. Paxton always anxious to advance himself, applied for a job there and at the age of twenty in 1823, was working at the centre of the gardening world. He had given his date of birth as 1801 instead of 1803. The Duke who was president of the Horticultural Society, used to walk into Chiswick Gardens. Paxton, who was by now a foreman responsible mainly for creepers and new plants, often opened the gate for him. The Duke talked to him and was impressed by his 'trim, manly, intelligent bearing'. Thus when the Duke needed a head gardener at his Chatsworth estate, Derbyshire, he offered the post to Paxton. It was a splendid opportunity which Paxton seized immediately. On a wage of 18s. a week at Chiswick, and about to emigrate to America, he now had the chance to serve one of the wealthiest aristocrats in the country at a wage of £70 a year and a cottage. He went down to Chatsworth straightaway. He described what happened:

'I left London by the Comet Coach for Chesterfield; arrived at Chatsworth at 4.30 a.m. in the morning of the ninth of May 1826

As no person was to be seen at that early hour, I got over the green-house gate by the old covered way, explored the pleasure grounds and looked round the outside of the house. I then went down to the kitchen gardens, scaled the outside wall and saw the whole of the place, set the men to work there at six-o'clock; then returned to Chatsworth and got Thomas Weldon to play me the water works and afterwards went to breakfast with poor dear Mrs. Gregory and her niece. The latter fell in love with me and I with her, and thus completed my first morning's work at Chatsworth before nine o'clock.'

Mrs. Gregory was the housekeeper and her niece, Sarah Bown became Mrs. Paxton eight months later. This marriage was very important for the whole of Paxton's career as well as for his happiness. His wife was the daughter of a typical man of the industrial revolution. He was a prosperous farmer who became a successful manufacturer of spindles for the new cotton factories of Derbyshire, and was able to give his daughter £5,000 when she married Paxton. Not only did Sarah Bown have money but she was also a very competent business woman. She was Paxton's assistant in carrying out his great plans at Chatsworth and also in his many business ventures later on. They had eight children.

At Chatsworth, between the ages of twenty-four and forty, Paxton became the foremost gardener-botanist in the country. He made himself so useful to the Duke of Devonshire that his responsibilities were increased. He found the gardens and pleasure grounds in poor shape but he soon put them in order. 'Twelve men with brooms in their hands on the lawns began to sweep, the labourers to work with activity', wrote the Duke. Paxton was made forester as well and then put in charge of the estate roads. Before long he became the Duke's agent or manager for the whole great estate.

At the same time his name became widely known in the horticultural world through his journalism. When he was twenty-eight he started, with a partner, the *Horticultural Register*, a monthly gardening magazine which ran for four years. The next venture, *Paxton's Magazine of Botany and Register of Flowering Plants* was more ambitious and ran for over ten years. Others such as the *Gardeners Chronicle* followed until he was nearly fifty, the most notable being the *Pocket Botanical Dictionary; comprising the Names, History and Culture of All Plants Known in Britain* which he published jointly with a professor in 1840. In this way Paxton gained a national reputation as a botanist as well as a gardener.

It was in these publications that Paxton described many of his achievements both with plants and, more important for the future, with building glasshouses. Many of the plants were brought from abroad; the Duke sent a gardener trained by Paxton to South Africa and India. Some, Paxton brought home from his travels in Europe and the near East with his master. After years at Chatsworth he had become the Duke's friend as well as his servant. In 1838 the two, with a retinue, made a year's grand tour of Switzerland, Italy, Malta, Greece and Turkey. Paxton was in charge of the party. He wrote frequently to his wife:

Oct. 2nd, 1838

'I arrived at Geneva about 7.30 last night and after some difficulty found the Duke. . . . His Grace has been in a terrible stew for want of me for some time. He would reserve some of the grand sights of Switzerland until I came, and they all say how he has worried himself. I am now to be grand leader of the band until all the grand sights are exhausted.'

Como, Oct. 14th, 1838

'The order of our march is this. The Duke and myself in the first carriage with Robert the footman on the box. The Doctor, Meynell and Weiler in the other carriage with Bland on the box. The courier rides before. The Duke dislikes the Courier much . . . he is the greatest liar I ever heard speak.'

Milan, Oct. 21st, 1838

'Pray tell Andrew that I am doing nothing at French now, but he will find on my return that I can speak Italian. The Duke gives me lessons as we go in the carriage. I am already beginning to speak a little. It is the most beautiful language in the world, full of music and sweetness.

Venice

'I will tell you how we manage our living affairs. The Duke pays for all coach hire, rooms, bed and attendance. We find our own eating and drinking. . . . Eating is cheap enough but wine that will not give you the belly ache is five francs a bottle. I manage better when we are travelling, for I dine with the Duke every day in the carriage. You will be surprised to hear how much I smoke.'

Rome, Jan. 3rd. 1839

'You will be astounded at my knowledge of sculpture, Pictures and other things belonging to the Fine Arts. . . . At this moment I believe I

know as much of Rome and its contents as any person in it. I have
bought the History of Ancient Rome and read it on the spot. . . .'

Naples, Feb. 9th, 1839

'If I had not you all at home wanting me, as well as so much business,
I should not care how long I remained out, for travelling is the most
delightful thing in the world, and more especially with such a nobleman
as the Duke, who has only to will to go where he pleases. His Grace
takes quite as much interest in showing me everything as he did when
we first started.'

Malta, April 2nd, 1838

'I should like something done with the old Mill at Matlock this season.
Pray see about getting plans for a couple of cottages to be built out of
the old materials. Have everything ready for me to decide when I
come home. . . . You don't say a word about the great Stove, but I
suppose it is going well.'

The 'great Stove' was the great conservatory, then being constructed
at Chatsfield, to Paxton's design, under his wife's watchful eye. She
wrote, 'the Conservatory is going on very well in another fortnight
the intermediate ribs will be made and they are now fixing things up
for glazing.' It was 277 feet long, 123 feet wide and 67 feet high, the
biggest in Europe.

By this time Paxton had been experimenting with the construction
of glass-houses for ten years. After several years he developed his ridge
and furrow type of roof. He described this in his *Magazine of Botany*
in 1825.

'About three years ago it occurred to us that wooden roofs would
admit much more light, if the sashes were fixed in angles. We tried a
small range of houses on this principle, with the sash bars fixed length-
ways, the usual way, and rafters to bear up the lights. These houses
were very light, and the plan appeared to possess several advantages—
1st., more morning and evening sun were received, and at an earlier
hour than a flat roof house; and 2ndly; the violence of the mid-day
sun's was mitigated by the disposition of the angled lights receiving
the sun's rays in an oblique direction. Subsequent experience has led
us to make several more alterations, such as doing away with rafters
altogether, changing the longitudinal positions of the sash bars, &c.
as will be seen in the annexed engravings.'

This was not a new idea but Paxton was the first to develop it and improve on it. The principle of 'sashes, fixed in angles' was used in all his later glass buildings. Two years later when building a large greenhouse he introduced other basic improvements. There were no outside gutters or down-pipes as the rainwater was carried away in the cast-iron columns supporting the wood and glass roof, which were hollow. Nor were there doors, but in their place the glass of the sides slid up and down in grooves. Later he designed his own type of gutter, a timber piece acting as both rafter and gutter, to carry away internal moisture and rain. Thus when he came to start the 'great Stove' in 1836, he had developed his own system of roofing to put into practice on a large scale. This great conservatory was a fore-runner of the Crystal Palace, fifteen years later.

It was constructed over five years and cost £33,099. 10s. 11d. The curvilinear structure was the most complicated Paxton had so far attempted. It was a wooden structure, sashes and ribs of wood, and iron was used only where it was unavoidable, for the thirty-six columns and the rectangular frame which they supported. Paxton explained twenty years later his preference for wood:

'If you have wooden bars and iron gutters you will have the greatest difficulty in the world in making them act together. The two do not act together. Metal expands and contracts and is subject to certain influences; wood is a non-conductor and is not affected in that way. . . . I should never bring any iron to exposure to the atmosphere. I think if you had at all an inclination to put up iron gutters, you should have our gutters at Chatsworth, which have been up twenty years. See how well they stand, where the water is carried off from the wood, because I maintain that the great principle to keep wood from rotting is to keep it exposed to the atmosphere. Wherever you keep wood covered up it will not last very long; for instance, if you lead these gutters they would not stand; they would rot three times as fast. You have only to take care that the water is quickly delivered from it, and to keep it well painted, and there is no end almost to the time that good wood would last'.

Forty miles of sash bar were required. Paxton invented a machine to make them, driven by a 3 h.p. steam engine from Boulton & Watt, which saved over a thousand pounds in labour. For the enormous quantity of glass he had to visit the firm of Chance in Birmingham which had recently started a new method of making sheet glass, and insist on having four-foot lengths. Glazing was a problem. As a contemporary writer put it:

'I must leave my readers to guess how a dozen or two painters and glaziers may be enabled to crawl spider-like, freely and nimbly, over a surface of such fragile materials, without either bending a single one of the slender ribs, or fracturing a pane of glass.'

Fuel for the eight boilers was brought by an underground tramway from underground stores. The great conservatory was a sensation for many years among the wealthy and influential and made Paxton's name.

During the next few years Paxton built a number of glass-houses. In these he was working his way to the best and most useful type of wood and glass structure, one which combined his ridge and furrow with flat roof and vertical sides. Eventually in 1849, two years before the Great Exhibition, he arrived at what he thought to be the best structure. This was the lily house for *Victoria Regia*.

This great lily had been brought from British Guiana to Kew Gardens, but it refused to flower. Paxton decided to try to make it flower at Chatsworth. Having obtained a plant from his friend the director at Kew at the beginning of August, he put it in a special tank in the great conservatory. By mid-September the leaves were 3 feet 6 inches in diameter: Paxton watched over it like a baby; he thought 'if Electric light was not so expensive I should use it for two or three hours morning and evening all winter'. That was before the days of dynamos. Early in November he announced to the Duke, 'Victoria has shown flower!! An enormous bud like a poppy head made its appearance yesterday. It looks like a large peach placed in a cup. No words can describe the beauty and grandeur of the plant.' The leaves were now nearly five feet in diameter. They were so strong that Paxton placed his daughter Anne, aged seven, on one. Soon the lily had to be moved into its new house.

The structure of the lily house owed a good deal to the structure of the plant. Paxton himself said, 'Nature was the engineer. Nature has provided the leaf with longitudinal and transverse girders and supports that I, borrowing from it, have adopted in this building'. The flat surface of the giant leaf was in fact supported by cantilever-like ribs radiating from the stem and secondary ribs which were held in place by cross members. The lily house, a rectangle, 61 foot 6 inches by 46 foot 9 inches, containing the main circular tank 33 feet in diameter, was made entirely of glass, wood and iron. The roof, on the ridge and furrow model, completely of wood and glass, with Paxton's own gutters, was supported by a frame of four wrought-iron girders which were in turn supported by eight cast-iron columns. These hollow columns were also drain pipes. The wrought-iron girders were connected by tension rods. The sides of the house were glazed lights on wooden sashes. The floor was slatted to act as a dust trap and a ventilator.

The Crystal Palace was virtually this on a much larger scale, and on eighteen acres instead of 300 square yards. It was designed a few months after the lily house was completed.

Paxton made the first sketch for the Crystal Palace on a sheet of blotting paper in the board room of the Midland Railway at Derby in June 1850. This shows that, at forty-six years of age, he had become a good deal more than the foremost gardener-botanist in England. He had for two years been a director of the railway company and on this famous occasion he was chairman of the Way and Works Committee which was trying a pointsman for a minor offence. The pointsman was fined five shillings, about a quarter of a week's wage, as was customary before the railwaymen had a trade union to protect them.

By this time Paxton was very well off. From his early thirties he and his wife had speculated in railway shares, and successfully. In that age of railway manias there were many opportunities of growing rich for the skilful speculator. The following letter of Paxton's shows a typical situation:

Adelphi Hotel, Liverpool
July 27, 1839

My Dear Love,

. . . I went into Liverpool Market this morning and during the panic on this railway bought ten more shares at £40 a share. This is the lowest sum any shares have ever been sold at. They have rallied this afternoon and are selling at £42. 10s. but if I can raise the wind in any way I will hold them all until the Bolton and Preston line is open, as I am sure it will pay then. But this was not the cause of my purchasing them. I saw from the manner of the reports, from the manner of the officers, and from the manner of the directors, that they were anxious to depress the value of the shares to the lowest possible value; intending, no doubt, to buy in largely now. . . . I am so positive of this from the manner of the whole of them that I would buy 100 shares to-morrow if I could find money to pay for them.'

He also promoted several small railways, sometimes in the Duke's interest, sometimes in his own; for example the Manchester, Buxton, Matlock and Midland Junction Railway. He was a close friend of George

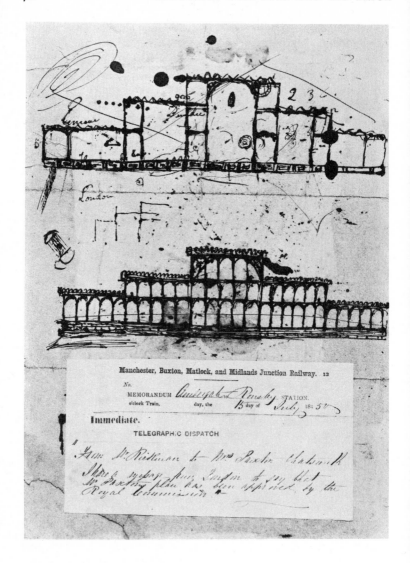

Paxton's original sketch for Crystal Palace

and Robert Stephenson and Thomas Brassey, the great contractor, and on friendly terms with George Hudson, the railway king.

In other directions, more connected with his job at Chatsworth, Paxton had also become well known by 1850. He had designed the layout of parks; Prince's Park at Liverpool, Upton Park at Slough and Birkenhead Park, for which he received £800. He had also designed houses round about the great estate, in particular the whole village of Edensor. Bolton Abbey in the West Riding, one of the Duke's properties, was altered according to Paxton's plans. He wrote to his wife:

'It will be a big job altogether and one that I am sure will do me great credit. I am already here at at Chatsworth, not a thing can move but I must be consulted and everybody pays me court and cringing to me which to tell the truth rather annoys me.'

Thus when Paxton designed the Crystal Palace he had already a high reputation and considerable wealth. He had been able to subscribe £25,000 to starting a newspaper, the *Daily News*.

In the competition for the building to house the Great Industrial Exhibition in Hyde Park 245 designs were entered. The building Committee of the Exhibition Commissioners rejected them all and decided to produce one itself. Paxton happened to visit the new House of Commons where a session was being held to test the acoustics of the new chamber. He could not hear a word and this prompted him to think that the Government was making another great mistake with the Exhibition building. 'I decided that I would prepare plans for a glass structure', he said, 'and the first thing I actually did was to go to Hyde Park and step over the ground to ascertain the length and breadth on which the Building was to stand'. A few days later he made his sketch, took it home and sat up all night working it out to his satisfaction, completed the drawings with his staff at Chatsworth in eight days and submitted them to the Commissioners. When these gentlemen could not reach a decision Paxton published his design in the *Illustrated London News*—in which the Commissioners' own design of an enormous brick building had also appeared—and the press welcomed it.

He now had to obtain a tender quickly. He was fortunate in getting a firm of contractors, Fox, Henderson & Co. of London, Smethwick and Renfrew, to collaborate fully. He said:

'When I first thought of sending in a design I had to consider not only what I knew myself but what would be thought practicable by others; not only the mechanical realisation of my design but also the possibility of its realisation within a given time. That I did not err in this last respect I owe to the ability, energy and transcendent skill of Messrs. Fox and Henderson. The structure was an entirely novel and new one to them; previously they had no experience in the peculiar plan of roofing and drainage which was for the first time brought before them in my plans for the roofing and therefore they had to rely entirely upon my experience in these matters.'

Charles Fox, the contractor, had been employed by Robert Stephenson on railway construction. He stepped out of his office in Portland Place and having measured that Place found the Crystal Palace would be nearly the same length, and as high as the houses but three times its width. After intensive work the tender was ready within a fortnight. It was for £150,000 or £79,800 if the materials remained the contractor's property when the building was taken down. By this time Paxton had made an important alteration to his plan. The commissioners wanted the building to include the trees on the site so as to avoid cutting them down. Paxton's solution to the problem was as follows:

'I went direct with Mr. Fox to his office, and while he arranged the ground plan so as to bring the trees into the centre of the Building, I was contriving how they were to be covered. At length I hit upon the plan of covering the Transept with a circular roof similar to that on the great conservatory at Chatsworth, and made a sketch of it, which was copied that night by one of the draughtsmen, in order that I might have it to show to Mr. Brunel, whom I had agreed to meet on the ground the next day. Before nine the next morning Mr. Brunel called at Devonshire House, and brought me the height of all the great trees.'

Just before the Commissioners accepted the tender on 26th July. Paxton applied for a patent to cover his system of roof construction, It described 'improvements in the construction of the description of roofs known as ridge and valley roofs, part of which improvements are also applicable to other descriptions of roofs.'

The speed of construction was what impressed contemporaries. The contractors had four months to enclose an area of eighteen acres and complete the job. They had undertaken to hand over the building by the last day of 1850 in order that the Exhibition might be opened on 1st May 1851. Fox later described the first steps:

'I set to work and made every important drawing of the Building as it now stands with my own hand . . . the Commissioners gave us possession of the ground on 30th July when we proceeded to take the necessary levels and surveys and to set out with great precision the position of the various parts of the building. . . . The drawings occupied me about 18 hours each day for seven weeks and as they came from my hand Mr. Henderson immediately prepared the ironwork and other materials required in the construction of the building. . . . As the drawings proceeded the calculations of strength were made and as soon as the particulars of the important parts were prepared, such as cast iron girders and wrought iron trusses we invited Mr. Cubitt to pay a visit to our works at Birmingham to witness a set of experiments in proof of the correctness of these calculations.'

The 'Mr. Cubitt' was William Cubitt (see Chapter one). He kept a watchful eye on construction on behalf of the Commissioners. Fox continued:

'The preparations of the iron work and other materials was pushed forward with the greatest vigour . . . so that on the 26th September we were enabled to fix the first column in its place. From this time I took the general management of the Building in my own charge and spent all my time upon the works feeling that unless the same person who made the drawings was always present to assign to each part as it arrived upon the ground its proper position in the structure it would be impossible to finish the Building in time. I am confident that if . . . as is usual in the construction of large buildings, the drawings had been produced by an architect and the works executed by a contractor instead . . . these separate functions being combined by my making drawings and then superintending the execution of the work; a building of such dimensions would not have been completed within a period considered by experienced persons altogether inadequate for the purpose.'

Paxton himself watched his design being realised:

'I was on the ground last Saturday and saw two columns (18 feet high) and three girders put up by two men in sixteen minutes. The whole of the columns were fitted together like the joints of a telescope formed in the same mould, and put in sockets ready to receive them, and then fastened with bolts. The admirable nicety with which one part was made to tally with the other.'

Before the columns were raised the foundations of concrete piers had been laid in August. On these were cast-iron plates with sockets into which the columns were fitted. There were 1,060 columns in each of the three tiers of the building. Being hollow they acted as drains for both rainwater and internal condensation. At the top the columns were joined by light strutted cast-iron girders three feet in depth.

Progress was rapid until at the end of November the roof was started. Paxton's own patent roof system was used throughout. There were 205 miles of his grooved sash bar and 24 miles of the Paxton gutter on the ridge and furrow roof. In the gutter, a solid beam of wood, a deep groove on the top received the rainwater from the glass roof while two smaller grooves on each side gathered the internal condensation; the water then ran down inside the columns, into underground cast-iron pipes and then to drains and sewers. Paxton described the system as follows:

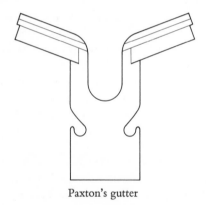

Paxton's gutter

'The gutters are arranged longitudinally and transversely; the rainwater passes from the longitudinal gutter into the transverse gutter over the girders, and is there conveyed to the hollow columns, and then to the drains below. As these transverse gutters are placed at only 24 feet apart and as there is a fall in the longitudinal gutters both ways, the water has only a distance of 12 feet before it descends into the transverse gutters which carry it off to the hollow columns or down pipes. The grooves for carrying off the moisture which condenses on the inside of the glass are cut out of the solid; in fact the whole gutter is formed by machinery at one cut. The gutter is cambered up by tension rods having screws fixed at the ends so as to adjust it to the greatest nicety, as is the case with the wrought iron girders which span the Victoria Lily House.'

The glass used came to a total of 900,000 square feet, in panes 49 inches by 12 inches, all supplied by Chance Brothers. The glazing of 17¾ acres of roof was carried out by means of small covered trollies the wheels of which ran along the deep grooves in the gutters. There might be half a dozen trolleys at one time, working backwards as the glazing progressed. In each trolley, two men and two boys; there was in fact a strike by the glaziers about piece-work rates.

In the following month, December, the flooring was laid. Paxton told how this developed from his experience with conservatories:

'Many experiments led me to the adoption of trellised wooden path-ways, with spaces between each board through which, on sweeping, the dust disappears. The boards for the floors will be 9 inches broad and 1½ inches thick, laid half an inch apart, on sleeper joists 9 inches deep and 3 inches thick placed 4 feet apart.'

He had also arranged for sprinkling machines.

On 8th December Fox began what he described as 'perhaps the most difficult and certainly the most hazardous portion of the work', and carried it out successfully. This was the raising of the sixteen great semi-circular ribs of the transept. They were all placed in position in eight working days by a special gang of longshoremen under the direction of Fox and Henderson themselves, watched by a crowd who had paid five shillings each to enter the enclosure. The Prince Consort, Prince Albert, watched the operation on 8th December and was loudly cheered by the 2,000 workers on the site.

The contractors were not able to hand over the building on 1st January 1851 as promised and were given another month. This was because an additional gallery round the second tier was required. The total cost exceeded the tender by £30,000 but this sum was paid to the contractors. It was an acknowledgement of their skill and speed. The exhibits began to pour in in February and the Exhibition was duly opened by Queen Victoria on May 1st, the appointed day.

The Crystal Palace was greatly admired by the visitors of whom there were six million; but there were also many fears about its safety because it was so novel. The sound from the organ pipes would shatter the glass which would in any case be broken by hailstones; the structure would not stand against winter gales; the galleries would collapse and crush all under them. But great precautions were taken. The galleries particularly were tested repeatedly—by hundreds of workers jumping up and down, by Sappers marching, doubling and marking time. There were no mishaps.

Paxton became famous. The Queen knighted him and the Commissioners gave him £5,000 from the profits of the Exhibition. She noted in her diary on 1st May: 'All the Commissioners, the Executive Committee etc., who worked so hard and to whom such praise is due, seemed truly happy, and none more so than Paxton, who may be justly proud; he rose from being a common gardener's boy.' Paxton himself explain his success:

'My reason for offering a design for this building was this: When the plans for the structure were sent in by various parties to the Royal Commission many forcible reasons were urged in the daily papers against the propriety of erecting a large building of bricks and mortar in Hyde Park.

'The system of ridge and furrow *flat* roof so fully in my mind, it only required adaptation of the principle on a large scale to suit a vast building for the Exhibition.

'It has been said that "it is a fortunate idea", but the idea though fortunate was not fortuitous. It was the result of long study and long labour. The great experience I had in the erection of glass structures and the invariable success which had attended my exertions emboldened me to produce a design. At that time I was erecting a house of peculiar structure which I had designed for the growth of that most remarkable plant Victoria Regia—it is to this plant and those circumstances that the Crystal Palace owes its direct origin. From the simplicity of all its parts, together with the simplicity of the detail the structure does not offer a subject for a long description; it is only by the multiplication of these parts that the stupendous structure now in progress is extended.'

It was the first large building made of prefabricated parts and assembled with mass production methods. *Punch* urged Paxton to design a glass Houses of Parliament as Barry's building was taking so long to finish.

After the Great Exhibition Sir Joseph Paxton, though still the Duke's agent, branched out into ever increasing business activities. He first organised the rebuilding of the Crystal Palace at Sydenham on a much grander scale. He became busy as an architect. His two most notable buildings were both for the wealthy Rothschilds: Mentmore mansion, park and village in Buckinghamshire (1850–1854) and Ferrières, near Paris (1854–1859), a vast country house with its five sets of private apartments for the family, eighteen suites for visitors and staterooms. At the same time he became Member of Parliament for Coventry as a Liberal in 1854, was easily re-elected twice and held the seat until the year of his death. Never a real politician, he made use of his special knowledge as a back-bench member and in committee. Thus during the Crimean War, as an M.P. he argued strongly for a special Army Works Corps of labourers and navvies to be sent out for road construction. Thanks to his organising, the first detachment duly arrived, a month before the fall of Sebastopol.

As an M.P. Paxton did his most important work for metropolitan improvements; a long and complicated project to improve and bring up to date the drainage and sanitation including the river Thames, in London. Two of his contributions took up much time during his last ten years. The first was his scheme for a great circular traffic way in and round London. In 1855 Parliament set up a committee 'to inquire into the State and Condition of the several Communications to and in the Metropolis, including the Bridges over the Thames, and the Approaches thereto; whether they are adequate to the present and increasing Traffic of the Metropolis; if not, the best mode of Improving the same'. Paxton's scheme for this was his Great Victorian Way. This was a two-tiered route of iron and glass running round London. The lower tier of roadway with shops and houses was under glass; the upper tier was to have eight lines of railway. Starting from the Royal Exchange, it was to cross the river near Southwark Bridge, recross it at Westminster, then go via Victoria, Belgravia, Kensington, Paddington, Islington and home. This great scheme was to cost thirty-four million pounds, and nothing more was heard of it. Five years later, Paxton was chairman of a House of Commons committee to enquire into the Thames Embankment, and it was largely due to his energy that London had the Victoria Embankment by 1870.

In addition, Paxton's ambition led him into more and more business projects. Already a railway director, he now turned to railway construction as an associate of the great contractors, Thomas Brassey and Morgan Peto (formerly of Grissel & Peto who had built most of the Houses of Parliament). With these and other businessmen he was responsible for railways in Spain, India, Turkey and Ireland. He also had money invested in railways in the United States, South America and Canada. Although his health began to give way he drove on with other projects: docks—he promoted the Thames Graving Dock; shipbuilding—he helped to finance Brunel's Great Eastern. Obliged by ill health to resign from Parliament he died, wealthy but worn out, aged sixty-two in 1865. *The Times* wrote:

'He rose from the ranks to be the greatest gardener of his time, the founder of a new style of architecture, and a man of genius, who devoted it to objects in the highest and noblest sense popular.'

SIR GEORGE GILBERT-SCOTT (1811–1878)

Unlike Paxton, Scott came from the middle class. His buildings belonged to the past while Paxton's pointed to the future. They were traditional in appearance, a deliberate revival in the nineteenth century of the Gothic style in which the cathedrals and churches had been built in medieval times. They were typical of many Victorian buildings.

Born at Gawcott, a small village near Buckingham, Scott was the third of four sons of the Church of England minister in that place. His grandfather, too, was a clergyman. His mother's family came from the West Indies where they had owned valuable plantations. The Scotts, however, had little money. The living was a poor one and there was no parsonage until the Reverend Scott raised money and designed one himself. He also designed a new church for the village. He took pupils who were entering the church in order to supplement his income. When they made their annual outing to Stowe, the great house of the Marquis of Buckingham, four miles away, the Revd. Scott rode his old horse, the older boys walked and Mrs. Scott and the young children rode in the baker's cart.

Scott never went to school but was taught at home by his father. He wrote:

'My eldest brother was a youth of remarkable talent and was viewed as a little god by his brothers and even by his parents. This had a bad effect on me . . . all efforts were concentrated upon him. His next brother got a little attention at second hand and being a boy of steady industry and good ability, he got on; but I, the third was too far removed to pick up even the crumbs. . . ! I ought certainly to have gone to school, but this was out of the question. My father was poor. . . . I do believe, however, that if encouraged and helped, I should have done well, and in mathematics I did get on fairly. My great relief . . . was the visit of the drawing master. . . . His visits twice a week were the very joy of my life.'

The drawing master taught him to sketch the neighbouring churches. He began to take a keen interest in these old churches and their different styles of architecture. When his father noticed this he decided he should be an architect and at the age of fourteen he was sent to his uncle at Latimers, near Chesham, for preliminary instruction. There he spent a happy year learning trigonometry, mechanics and drawing. He was always a lonely child without friends of his own age, whose father did not think much of his ability.

When he was fifteen Scott was apprenticed for four years to a little-known architect in London. He learned thoroughly the routines of building and specifications but he was discouraged from indulging his fancy for Gothic design. His father was told that he was wasting his time in sketching medieval buildings. Having completed his articles he spent a year with the contractors, Grissell & Peto, 'giving such services as I could offer, in return for having the run of their workshops, and of their London works'.

This experience of practical work was invaluable. Grissell & Peto were then building Hungerford Market, later demolished to make way for Charing Cross station. Scott wrote:

'The work was constructed on principles quite new. Iron girders, Yorkshire landings, roofs and platforms of tiles in cement and columns of granite being its leading elements. I got much information too in the joiner's shop, from the foreman, from the clerks in the office, and especially from assisting in measuring up work, usually with the foreman. . . . Mr. Peto did not quite relish my prying so closely as I was wont, into the foundations of the prices of work and materials. . . .'

Finding it necessary to earn a salary, at the age of twenty-one, he became a clerk with another architect and helped him with the working drawings and measuring up of the Fishmonger's Hall which was being built by Cubitts in Lower Thames Street. During his two years' stay there he acted as clerk of works on a small job and also designed two houses on his own account. Then he started his own practice, living in chambers in Regent Street. He soon found plenty of work, of a kind typical of the age. This work came as a result of the institution of the Poor Law.

In 1834 the Poor Law Amendment Act tightened up on the poor. Its aim was to economise by discouraging the poor from applying for relief. Accordingly relief, in cash or kind, outside the workhouse was forbidden. Parishes were formed into unions each of which was to

have its own workhouse. These workhouses were to be a deterrent to
the poor; the conditions in them were designed to be worse than what
the lowest paid labourer could get for himself. The poor hated the
workhouses; they called them the 'Bastilles'. Many new workhouses
were built from 1834 onward throughout the country. They launched
Scott on his career.

A friend of Scott, 'a young and inexperienced architect' called
Kempthorne had been asked, through his father's influence, to design
model plans for the workhouses. Scott joined him for a short time
and then, using this experience, canvassed for work on his own account
from the boards of guardians in the unions where his father had been
known. Thus began a successful business in which he took a builder's
son as partner. He wrote:

'For weeks I almost lived on horseback, canvassing newly formed
unions. Then alternated periods of close, hard work in my little office,
with coach journeys, chiefly by night, followed by meetings with
guardians, searching out of materials, and hurrying from union to
union, often riding across unknown bits of country after dark. . . . I
built, I think, at that time two union houses in Bucks, five in North-
amptonshire and four in Lincolnshire.'

The partnership built, in fact, some fifty 'Bastilles' by 1845. On the
basis of this prosperity when he was twenty-six Scott married. Workhouses were all right for bread and butter but would not make his
name. What he wanted to do was design new churches and restore old
ones. In 1838 he competed successfully for his first church, at Lincoln.
Two years later he was invited to compete for the Martyrs' Memorial
at Oxford, thanks to the influence of friends, and won. Thereafter his
church practice grew rapidly. He became well known as an expert at
restoration of churches and cathedrals. Ely cathedral appointed him
restorer in 1847, followed by Westminster Abbey and a dozen other
cathedrals. Later there was much criticism that he restored too much
and thus spoilt the appearance of the buildings.

When England, together with France and Sardinia, had defeated
Russia in the Crimean War the Government decided to build a new
Foreign Office and a new India Office in Whitehall. Scott's design won
the competition and he was appointed architect for both. At this
time, however, there was a great controversy about the merits of the
Gothic or vertical pointed style versus the classical horizontal style for
public buildings. Scott was the leader of what was called the Gothic

revival to which we owe the appearance of many Victorian buildings today. His design was an elaborate one similar in general appearance to St. Pancras Hotel today. The question was debated in Parliament. Many members thought it was wrong and not dignified enough for the great buildings at the heart of the Empire. Others such as Joseph Paxton and Morton Peto supported Scott. Unfortunately for Scott, the prime minister, Lord Palmerston, disliked his design. Scott wrote:

'Lord Palmerston sent for me and told me in a jaunty way that he could have nothing to do with this Gothic style, and though he did not want to disturb my appointment, he must insist on me making a design in the Italian style, which he felt I could do quite as well as the other. That he heard I was so tremendously successful in the Gothic style, that if he let me alone I should Gothicize the whole country, etc., etc.'

Scott gave way and designed the buildings as they now stand, very reluctantly, but the appointment was too important to give up.

The building for which Scott is best known, and which he thought was his best is the Albert Memorial in Kensington Gardens. He was invited to compete for it after the death of the Prince Consort in 1861. This elaborate structure built in the form of a shrine, cost £120,000. The materials used included marble, granite, Portland stone, enamels, precious metals and plentiful gilding. His designs in a similar style for the Albert Hall were not accepted. It was eventually built by two engineers.

The second building for which Scott is well known is also in London. It is the St. Pancras Hotel fronting the main line terminus, on the Euston Road, close to Lewis Cubitt's King's Cross Station. Scott was by now the most famous architect in Britain. He had a very large practice with many employees. But a commercial project like a hotel was new for him. He was persuaded into competing for it by one of the directors of the Midland Railway who was a friend of his. His motive for competing was, in his own words:

'I made my design while detained for several weeks with Mrs. Scott by the severe illness of our son Alwyne, at a small seaside hotel at Hayling in September and October, 1865. I completely worked out the whole design then, and made elevations to a large scale with details. It was in the same style which I had almost originated several years earlier, for the government offices. . . . Having been disappointed, through Lord Palmerston, of my ardent hopes of carrying out my style in the Government offices, I was glad to be able to erect one building in that style in London.'

The railway had required 150 bedrooms. Scott's design gave 250 and as a result his estimate of £316,000 was the highest by £50,000. All the same the railway preferred his design. The reason was a simple one. It needed a big showy building at its terminus to establish itself in London. The plan of the building was a clever use of the site. The main block with its clock tower at the east end and the big archway entrance at the west, was set back some distance from the Euston road. This was to allow for a wide approach ramp in front and parallel return to the road. Then, starting at the big archway, the west wing curved round in a segment to the hotel entrance at its end on the road. At the rear of the main block the great roof of the train shed came up close to it but was separated from it by a screen to protect hotel guests from noise and dirt.

It was an age of building railway termini in London. Euston in 1837 was followed by King's Cross and Paddington in the 'fifties, and by Victoria, Charing Cross and Cannon Street in the 'sixties. Nearly all had grand ornate hotels in front of the train sheds. The Midland Railway wanted a terminus of its own and just as grand. Starting in 1844 that railway became in twenty years the most prosperous in the country. Essentially a provincial company, centred in Derby it had to use the Great Northern line to King's Cross for many years for access to London. Now the directors, with ample funds, were determined, after long consideration of the heavy cost, to have their own line into London. Keen businessmen, faced with sharp competition, they meant to have a hotel at the entrance to their terminus which would outshine all its rivals particularly Euston and King's Cross on either side. It was a commercial decision as well as a piece of showmanship. The chairman of the Midland Railway stated that it had been decided to have a hotel at St. Pancras so as to put it 'upon a par with the Great Northern, the London and North Western, and the Great Western'.

The necessary Act of Parliament authorising the Midland Railway extensions to London had been passed in June 1863. The line was to run 49¾ miles from Bedford via Luton, St. Albans and Hendon to St., Pancras. The cost was to be £2,333,330 but actually exceeded £5 million. Work began at Kentish Town in the autumn of 1864. The line had to cross the St. Pancras ancient burial ground and the corpses had to be reburied. On behalf of the Bishop of London this work was supervised by a young architectual assistant, Thomas Hardy, the novelist and poet. The site of St. Pancras station was cleared in the first half of 1866. In this phase the people living in seven streets were made homeless as their homes were demolished without compensation after the landlord had sold out to the railway. They fought their case in the courts but there was no compensation for these weekly tenants. This was only a part of the homelessness caused. In the demolitions for the line to London and for the whole terminus up to 4,000 homes in which 30,000 people lived, were involved. These were the families of working men.

Work on the station itself began in July 1866. The main problem was the great roof of the train shed. The first section of the first cast-iron rib was not hoisted until November 1867. The contractors, the Butterley Company, were having difficulty with the sheer size of the roof and in providing the great timber scaffolding which moved on rails inside the roof as it was constructed. It was designed by W. H. Barlow, one of the leading engineers of the time, who had been trained by Robert Stephenson and was a Fellow of the Royal Society.

The roof itself was, and is, in one great span of 240 feet, by far the largest span constructed at that time. Its special feature was that it was not completely round but had a pointed crown, in order to resist wind pressures. It had a Gothic shape, though for engineering reasons not architectural ones. Scott noted the point in his hotel design. Eventually on 1st October 1868, the first trains ran into the station.

The building of the hotel with which went the station offices, got off to a bad start. There was a serious financial crisis in 1866. In May the big finance house of Overend Gurney went bankrupt. This was followed by many other bankruptcies including several banks, two railway companies—the London, Chatham & Dover and the Great Eastern, and the great contractor, Sir Samuel Morton Peto (formerly of Grissell & Peto). Hence the directors of the Midland Railway not only delayed the start of the hotel but told Scott to economise. He reduced the cost by £20,000. He also had to agree to use bricks which were cheaper than those he had specified, although in fact the best bricks were used because the cheaper ones were not available. A second time the directors told Scott to 'reduce the cost of decoration and especially to dispense with the use of granite columns wherever he could do so without sacrificing the essential nature of the building'. At last in December 1867 they made a decision and put out phase one of the building to tender. This was for the work up to the first floor, on the eastern side of the great main entrance. The firms tendering included Cubitts and George Trollope & Sons but the contract went to another firm with a lower figure, £37,500, and work started in March 1868.

Scott himself had just received £4,000 as first instalment of his fee of 5% of the cost. He was anxious to maintain the very highest standards in construction but during that summer he had to use some inferior bricks and stone from a quarry which was not his first choice, There were delays in deliveries of bricks and stone but the directors wanted the work done quickly. Scott's clerk of the works, who was paid four guineas a week, who naturally followed Scott's ideas, reported 'under the circumstances I do not see how it is possible to do much more and ensure good work, there is always a great deal more difficulty in obtaining good work when so much pressure is put on. I have felt extremely anxious that they should not advance so rapidly as to abandon all hope in respect to quality and complete finish by substituting bulk and quantity'.

The wages currently paid to the men working on the station and hotel were as follows:

	per day
Labourers	4s. 6d.
Bricklayers, masons, carpenters	7s. 6d.
Smiths	8s. 6d.
Fitters	8s. 6d.
Boys	4s. od.
Plumbers	8s. 4d.
Slaters & glaziers	7s. 11d.
Riveters	8s. od.

By midsummer 1869 the second stage of construction was to be seen. It was built by the same contractors as before, at the same prices, and was not put out to tender. This stage included the clock tower at the eastern end of the front. In the autumn a journalist remarked on the 'expensively bright red brick Hotel'. He did not like what he saw because of the 'heavy looking walling and parapet in front of the hotel with the steep roadway to the offices' and he pitied the cab horses which would have to toil up it. In 1870 the builder's contract was extended to build the tower with its pinnacles and high-pitched roof, over the big archway where the straight front joins the curved west wing.

After that progress was slow and it seemed doubtful how far the building would be finished. However, in 1872 the railway directors appointed a hotel manager who accepted the post only on condition that the hotel was completed as planned. They, therefore, instructed Scott to press on with the whole hotel, although even then only two storeys of the curved wing were to be built. The same contractors continued, but now their prices were 10% more owing to the recent increase in prices of materials.

Scott was also involved in the furnishing and equipping of the hotel for which tenders were invited. He supervised the designs for furniture, for which he was paid extra. He actually designed four easy chairs himself and also 'the coloured ceiling decorations to some of the best bedrooms'. At last, on May 5th, 1873, the Midland Grand Hotel, as it was called was opened. Public opinion was very favourable, 'the apartments are reported to be magnificent and the charges moderate'.

The west wing of the hotel had still to be finished. Before that happened Scott was often in conflict with the railway directors. They wanted to control expenditure more closely whereas Scott wanted freedom to carry out his ambitious design. They rejected his design for the clock face on the eastern tower because of the cost. They told him he must not spend any money outside the contracts. Worst of all, when it came to filling the niches on the front with stone figures as planned they refused to agree to the expenditure. Scott, a confident man who expected to get his own way, accepted mildly all these set-backs. Provided he could have his elaborate, richly decorated, Gothic hotel in London he was satisfied.

The whole hotel was eventually finished in 1876, eight years after being started. It was certainly an impressive building. Looked at from the Euston Road, the clock tower at the eastern end with its oriel windows, rose to a height of 270 feet at the top of the spire. The main block, six storeys high, extended westward from the clock tower; a short distance along it was the gateway arch for arrivals in London. The ground floor of this block was occupied by offices, public waiting rooms, refreshment and dining rooms and cloakrooms, etc. From the first floor upward the block belonged to the hotel. There was a dining room 50 feet by 23 feet, smoking room, ladies' coffee room, ladies' reading room. On the first floor there were a dozen suites of rooms each having a bedroom, private sitting room, bathroom and dressing room. On the upper floors, 150 bedrooms. At the western end of this main block rose the other tower, 250 feet tall, and under it the second, larger, gateway arch for departures from London. From this tower the west wing curved forward to the elaborate portico and main entrance to the hotel. On the ground floor was the grand coffee room, curved in shape, 100 feet long by 26 feet wide by 26 feet high, bar, dining room for parties, and at the rear, billiard and smoking rooms, lavatories, laundry, drying rooms. The kitchen, etc., wash houses and engine room were in the basement. From the ground floor, the 'grand staircase' swept up to the first floor where the main feature was a ladies' reading room and dining room; more billiard rooms were at the rear. This floor had also fifty sitting and bedrooms. The second, third and fourth floors each had 70 rooms. The whole area of the hotel building was 2460 square yards, containing 500 rooms.

The construction was described in the *Building News*:

'The building is executed and faced with Nottingham Patent Brick Company's and Messrs. Tucker & Sons' (Loughborough, Leics.) pressed bricks which are, on the whole, very favourable in colour and texture. The bricks for the arches and quoins are from Messrs. Wheeler Bros, Reading. Ancaster stone is used for the tracrey, arches, labels and inside work, Ketton being employed for the weather parts. Red Mansfield is used in those portions where strength is required, as in the piers and bases. The granite shafts are of green and red Peterhead, and a large quantity of Shap granite is used. The construction of the floors is a system, known as Messrs. Moreland's patent system, of lattice girders covered with corrugated iron and Barrow lime concrete. The slates are green and present a pleasing contrast with the red bricks. . . . The warming apparatus is partly by warm air flues, in halls etc., and by open fires. . . . Ventilators of the 'hit and miss' principle are provided in some of the large rooms. The internal decorations are . . . partly from the designs of Sir George Gilbert Scott and are remarkably chaste as a rule; but we think the ladies' coffee room rather coarse, if not "loud".'

The fabric cost £304,335, decorations and fittings £49,000, furnishing £84,000, a total cost of £438,000.

The Midland Grand was profitable in its early years at least. It was expensive at 14s. for bed, breakfast and dinner, but it had a high reputation as a luxury hotel. Subsequently although other newer hotels overtook it in luxury it was always regarded as one of the best railway hotels. So it continued until 1935 when the London Midland & Scottish Railway converted it into offices because it was too difficult to adapt it to modern hotel standards.

Scott's account with the railway directors was settled two years before he died in 1878. He had to stand much criticism of his building.

People were either strongly for or against it. The critics expressed themselves strongly:

'The building inside and out is covered with ornament, and there is polished marble enough to furnish a cathedral. The very parapet of the cab road is panelled and perforated, at a cost that would have supplied foot-warmers to all the trains for years to come. . . . Showy and expensive, it will, for the present, be a striking contrast with its adjoining neighbour (King's Cross). . . . An elaboration that might be suitable for a Chapter-house, or a Cathedral choir, is used as an "advertising medium" for barmen's bedrooms and the costly discomforts of a terminus hotel. . . . everything looks out of place, and most of all the Gothic mouldings and brickwork, borrowed from the domestic architecture of the Middle Ages, which with its pretty littlenesses thrusts itself between the gigantic iron ribs of the roof. Add to all this the curious clumsiness of the medieval timbering of the roof of the booking office, in daring contrast with all the refinements of nineteenth century construction in the neighbouring shed. . . .'

Another writer called the hotel, 'Scott's tawdry masterpiece'; others, however, praised its 'convenience and its inspiring effect', and 'palatial beauty, comfort and convenience'.

Since then public opinion has swung to and fro. Before the second world war the hotel was condemned as the worst type of Victorian building, although one critic admitted that 'viewed in all the glory of the Euston Road it has a strange attraction, its Harz mountain top scenery and pinnacled summits point the moral of an active and prosperous career'. On the practical side too it was condemned. Four years after the hotel was closed the president of the railway explained '. . . it is impossible to put in a new piece of heating apparatus or anything of that kind without meeting with the same obstacles that would be encountered in modifying the Rock of Gibraltar'. Since the war, however, people have said it should be preserved as a monument to the Victorian Age. In 1967 the Government did take steps to preserve it by putting it on the list of buildings of special architectural or historic interest. The question is whether the hotel and station can remain a useful building as well as a national monument.

Scott, who had been knighted for his work on the Albert Memorial and at Windsor, was extremely busy with restoration of ancient buildings until he died. His very large practice was run almost on the lines of a factory. In his forties, before he built the Albert Memorial, his office, a three-storey house, was crammed with twenty-seven pupils, assistants and clerks. Pupils paid a fee of 300 guineas. Scott was too busy to look at all their work or to direct personally such a large staff. His practice was as much a commercial business as a professional one. Altogether he worked on over 700 buildings or projects including 29 cathedrals, 10 ministers, 476 churches, 25 schools, 23 parsonages, 58 monumental structures, 25 colleges or college chapels, 26 public buildings and 43 mansions. Two of his five sons continued this work.

A few months before he died he wrote this typical letter:

'Courtfield House,
Collingham Road,
S.W.
Jan. 22. 1878.

'I only last evening heard to my dismay that a movement was being made to induce our ground landlord to forbid the gardens of Courtfield Gardens being used by the more youthful inhabitants for lawn tennis.

'As holder of the leases of two of the houses I take the liberty of expressing to you in the most emphatic manner I am able, my heartfelt dissent from such a proposal.

'I seldom enter the gardens . . . but one of my greatest pleasures is to see how thoroughly they conduce to the happiness and enjoyment of the younger inhabitants, whose innocent and elegant disports it is always a solace to witness. . . .'

FURTHER READING:

George F. Chadwick, *The Works of Sir Joseph Paxton*, The Architectual Press 1961.
C. Hobhouse, *1851 and the Crystal Palace*, Murray 1950.
Violet R. Markham, *Paxton and the Bachelor Duke*, Hodder & Stoughton, 1935.
G. G. Scott, *Personal and Professional Recollections*, Sampson Low 1879.
J. Simmons, *St. Pancras Station*, Allen & Unwin 1968.

4 Building the Suburbs in Victorian London

BUILDING IN KENSINGTON AND CAMBERWELL

The Cubitts built houses, as is shown in Chapter 1, but most of these were grand houses in Belgravia and Bloomsbury for the wealthy middle class and the aristocracy. This chapter is about house building for the less well off, but still comfortably off, the 'respectable' middle class and lower middle class. These were the thousands of clerks, in commerce, industry, banking, insurance, railways, whose numbers had increased with the growth of Britain as a manufacturing nation trading with the whole world, and shopkeepers and retailers, small manufacturers, teachers, foremen, prosperous mechanics. For them were built the miles and miles of suburbs around London. In one such new suburb, Camberwell, 20,000 clerks lived. In the ten years from 1871 its population increased by 68%. On all sides of London the new suburbs spread out into the fields at Leyton, Ilford, Walthamstow, Edmonton, Wimbledon, Acton, East and West Ham, Kensington This vast development was made possible by the new railway suburban services, the horse buses, and later the trams.

This was the great opportunity for the speculative builders.

'The richest crop for any field
is a crop of bricks for it to yield
the richest crop that it can grow
is a crop of houses in a row.'

Most of these builders had only a small business, building not more than six houses a year. In the last quarter of the nineteenth century about a third of all house builders in London built only one or two houses a year. Even this they could do only in good years when interest rates on loans were low. In the bad years they kept going with jobbing. Anyone could be a speculative builder. Often he was a bricklayer, mason, carpenter, joiner, plumber or plasterer who took the opportunity of the expanding market for homes. He would lease a small plot of building land and put up a terrace of small houses, sub-contracting that part of the work which he could not do himself to men in the other trades. But he often went bankrupt. Others had no experience or skill at building. They were materials merchants or building societies who had to finish houses in order to recover debts. Other people became speculative builders by investing their savings; these included shopkeepers, merchants, tailors, domestic servants, publicans, mechanics, and quite often solicitors.

The speculative builder needed hardly any capital of his own to start, nor did he employ men directly. He borrowed money in various ways, from the landowner who had given the lease, on mortgage from building societies, from insurance companies, and from solicitors. He also got credit from the brick maker and timber merchant. If he was lucky or clever enough he could mortgage his houses one by one or even floor by floor and transfer the mortgages to the buyers of the houses provided he found them at the right time.

Things were made easy for the speculative builder who had no building knowledge by the numerous books, textbooks and manuals which were best sellers. There was the *Erection of Dwelling Houses, or The Builder's Comprehensive Directory, explained by a Perspective View, Plans, Elevations, and Sections of a Pair of Semi-detached Villas. . . . to which is added the specification, Quantities, and Estimate . . . calculated to render Assistance to the Young Artizan in every Department of the Building Act (1860); Rudiments of the Art of Building* which went through thirteen editions by 1870; and the *Builders' Practical Director of Buildings of all Classes making every Freeholder to be his own Builder, with Plans, Elevations and Sections for the Erection of Cottages, Villas, Farm Buildings . . . with detailed Estimates, Quantities, Prices etc.,* (1855), and so on.

The suburbs were for the most part then the work of small builders, but towards the end of the nineteenth century there were more large house builders. They could survive the depressions in the trade much better. The building industry was always subject to a cycle of boom and depression. When it was more profitable to invest money overseas, in the Empire, there was less left for building at home. As a result of one boom in building around 1880 there were many more medium and large sized firms and these could withstand the following depression. When the Camberwell–Peckham–Dulwich area of South London was being built up the three largest firms were each building twenty-five to thirty houses a year for a decade. These firms and a dozen like them did not build whole estates but a street or part of a street in one place and then in another, scattered through the area.

There were, however, some speculative builders, who concentrated their work on a simple estate. Here are two examples.

After the District Railway was extended to North End, Fulham, in 1873, West Kensington sprang up. The *Illustrated London News* reported in 1884:

'. . . speculative builders had money or credit; the tall houses, detached or semi-detached, or in closed lines improperly called terraces which ultimately became the sides of streets, rose up in a few months, roofed and windowed and calling for tenants.'

The building firm largely responsible was Gibbs & Flew, the largest one in West London and by that year it could tell a remarkable story of success. Established as a partnership only eight years before, it was turned into a limited company in 1882 with a capital of £100,000. This was increased to £250,000 in 1883. For that year the net profit was £47,365, a dividend of 7% was paid and £25,000 was put to reserve. Seven per cent was again paid in the next year.

The firm's profits came from three sources. It bought freehold land, made roads, and sold or leased it as improved building land in plots. This was the most profitable activity. It also built houses both for sale and to let. These were built to high standards of workmanship and with all the latest fittings, and went well. They had 'hot and cold water, and bathrooms with electric bells; while the encaustic tiles, stained glass and marble fenders gave them an attractive appearance not often found in houses of this class'. Prices were moderate because the firm made fittings in its own workshops and reduced costs by buying materials in bulk.

Investors were given the following reasons why they should put their money in the firm's shares:

'1. Messrs. Gibbs and Flew have purchased their own freehold estates, and possessing all the appliances and organisation for developing the properties, have not only made the builder's profits on the houses, but also the profits on the land and ground rents, which are generally the great source of profit and usually realised by the Freeholder.

2. The estates have been judiciously selected, all being in the Western and South-Western districts which are gradually increasing in value, by the natural growth of London in that direction.

3. The whole of the estates are on gravel soil, and are well drained.

4. The houses erected have been of moderate rentals (ranging from £30 to £100 p.a.), well built, and adapted to the requirements of the neighbourhood, in consequence of which there has been a continuously increasing demand for them.

5. Messrs. Gibbs & Flew own their own brickfields, extensive steam saw mills, joinery, stone and marble works.'

Three years after the limited company had been formed a curious transaction was carried out. The name of Gibbs & Flew Ltd. was changed to the West Kensington Estates Company, and Mr. Gibbs and Mr. Flew bought the whole of its plant, machinery and building business for £70,000. Messrs. Gibbs & Flew then carried on the business of builders as a partnership as in the first place. They advertised their business as follows:

MESSRS. GIBBS & FLEW

HAVING taken over the Building Business from the West Kensington Estates Company, Limited, with the several offices and works are now prepared to LET or SELL Freehold Building Plots upon the following Estates for the erection of houses, at varying rentals, as under:—

	RENTALS.	
The Cedars Estate, West Kensington....	£75	£150
,, Morning Park Estate, ditto........	60	100
,, Munster Park Estate, Fulham......	45	60
,, Baron's Court Estate, ditto........	60	100
,, Waldegrave Park Estate..........	70	100
,, Salisbury Estate, Fulham..........	35	45

MESSRS. GIBBS and FLEW have also a few Houses, ready for letting, on the above Estates, at the rentals quoted. These houses have been recently built upon the most approved design, and constructed with every convenience and accommodation, and will be decorated in a superior manner to the requirements of tenants, who may have the choice of papers, &c.

MESSRS. GIBBS and FLEW have also a few smaller Houses, suited to the requirements of genteel residents of limited means, which will be let at weekly rentals of 12s. 0d. to 14s. 6d. (according to size), to include all rates and taxes, &c.

On the other side of London, south of the Thames, Edward Yates was building an estate in Camberwell. The parish of Camberwell, including Peckham and Dulwich, had been a market garden for London since the sixteenth century. By the end of the eighteenth century it was already a place where citizens of London went to live to escape from the pollution of the city. When the Americans revolted against Britain there were still plenty of cow keepers and piggeries but according to a contemporary 'the spirit of building, which has been so prevalent for some years past; appears to have affected this part with any around the metropolis; for between Newington Butts and Camberwell, several new streets have been formed and a prodigious number of buildings erected'. Better communications accelerated this development. The new bridges across the Thames at Blackfriars (1796), Vauxhall (1816), Waterloo (1817) and Southwark (1819) made commuting to the City possible.

Consequently the population increased so that when Victoria's reign began in 1837 it was 35,000, But the rapid increase came later and when her reign ended the population had risen to 260,000—more than seven times. The fastest increase was 68 % between 1870 and 1880. What made this new suburb possible was first the London General or Thomas Tilling omnibus which had brought down the fare between Camberwell and the City to fourpence, the London, Brighton & South Coast Railway and the London Chatham and Dover Railway suburban services, the coming of workmen's fares and trains, and finally the trams of the Pimlico, Peckham and Greenwich Street Tramways Company. Thus thousands of people from the workman to the City banker could live in Camberwell and make the daily journey to work in London.

A great number of speculative builders provided the houses for them. At the peak of building work in about 1880 there were 416 firms or individual builders constructing 5,670 houses. They ranged in size from the one-man business building one or two houses to the large firm building more than sixty. Edward Yates was in the second class. The growth of his business over thirty years can be traced. Yates came to London from the north of England about 1850 without any money. He started by digging foundations for army barracks at Aldershot.

He began as a builder in 1867 when he leased part of an estate from another builder in Lambeth and built a dozen houses, borrowing money from the builder and a building society for working capital. These were evidently successful for in the following two years he built his first estate in Camberwell. This was quite a small one, Dragon Road, consisting of forty-six two-storey terrace houses. They cost about £190 each. Yates leased the land from the owner at low ground rents. He got his working capital in mortgages from three building societies, a solicitor and an insurance company, to the extent of about £7,000, on which he paid at least 5% interest.

After six years in Walworth where he built about seventy houses, north of Camberwell, he began his second project in the parish in 1875. This was a cul-de-sac of twenty terraced houses called Domville Grove. He again leased the land from the same landowner for seventy-five years at a ground rent equivalent to £5 per house. His working capital for this speculation he got by mortgages on the security of his earlier houses which had not yet been sold or had been let. In turn, the houses at Domville Grove while they were still his property, gave the security for more loans totalling £2,900 at 5% interest.

This cash enabled him to go on immediately to his third and much bigger speculation. In 1877 he bought nineteen acres of land freehold for £6,300, of which he paid £1,300 down and had the rest on mortgage at 4½%. The land was on the east and south sides of Nunhead Cemetery. Seven years later he had bought another 23½ acres for £13,250 and leased two other pieces of land for £225 p.a. ground rent adjoining his first purchase. By this time he had paid off the mortgage on the first purchase but had raised another mortgage of £4,000 on his subsequent purchases. This in turn, he paid off in two years. Thus he had an estate of some fifty acres, of which he had bought 42½ acres for £19,550 and leased the rest. In developing it and treading successfully the financial tight-rope, he had his other interests outside Camberwell in Walworth and Kensington to fall back on.

Yates began building the main road, Ivydale Road, between the Nunhead cemetery on the south and the railway line on the north, in 1884. By 1907 he had built 742 houses in a dozen roads on this estate. Before that time the estate had been provided with some of the basic facilities for living—small shops over which Yates had some control, a public house, a school, a church and a chapel. The houses themselves, which were mostly let, were good, solid buildings. Yates was under covenant to build houses of the value of £350 to £450 on that part of the estate which he had on lease. The rents were about £28 a year, excluding rates, less than those of a similar kind in the district. Their bay windows and stained glass and their 'back boilers, gas stoves, mahogany glass-fronted bookcases, numerous fitted cupboards, Venetian blinds, bathrooms, and separate washhouses' (H. J. Dyos, *Victorian*

Suburb) marked them off as respectable lower middle class homes.

The basis of Yates's success was his ability as a manager and organiser of building operations at different stages of completion on a large estate. He had to see that access to the site, materials, labour and capital were all available at the right time and the right place. Good access was important for reducing costs. At first Yates made his own roads but then they were sub-contracted and later they were made by the local authority. He also sub-contracted the sewers. Equally important were water and rail transport of materials. He was fortunate with water transport. The Surrey Canal ran from Surrey Commercial Docks, across Bermondsey, under the Old Kent Road and nearly to the Camberwell Road. Yates had bulk supplies of bricks and tiles brought by barge from Kent up the canal. However, this had to be hauled uphill by road from that point, a distance of some two miles. He therefore turned to the railway company and tried to persuade it to build a small private siding close to the site, but without success. However, he did get a better road access to Nunhead station which was on his doorstep. Later he made another unsuccessful attempt to get the railway company to build a new station near to the southern end of his estate, between Nunhead and Honor Oak stations.

His site organisation was efficient. According to H. J. Dyos:

'His almost legendary personal inspections of the progress of the works, the establishment of an effective site office complete with private telephone line to head office, and vigorous complaints to all suppliers of building materials who gave short weight or supplied inferior goods were Yates' means of maintaining his relatively high building standards.'

Labour did not apparently pose any problems for Yates. In spite of the growth of such trade unions as the Amalgamated Society of Carpenters and Joiners he did not pay union rates. And although the unions strongly opposed piece rates he seems to have paid them to the craftsmen he employed. The worker having agreed a rate was paid weekly an amount which roughly covered the work he had done and adjustments were made over longer periods of several months. H. J. Dyos gives an example of this:

'A carpenter named White, for example, contracted between April 1884 and May 1885 to prepare and fix some 1192 pairs of window sashes and frames (which, with Venetian frame fitting, was an almost unvarying employment for him for at least five years) to Nos. 2–200(e) Ivydale Road at a rate of 5s. 0d. reduced to 4s. 5d. per pair: his total

earnings for this period, less £2. 15s. for the "use of a machine," were
£292. 18s. 4d. of which he received in weekly instalments £281. 18s.
4d. and a final settlement in September 1885 of £11. When in work,
which was practically continuous, White was receiving on an average
about £5 a week. Tradesmen like him frequently employed their
own labourers.'

The final result was an estate of several hundred houses which
attracted tenants at reasonable rents. Yates himself prospered. When
he died in 1907 he was a millionaire. He had built altogether over 2,500
houses in south London.

There was another side to this picture of house building. The word
jerry-building came into use at the same period as the big expansion of
suburban building. How much jerry-building there was it is impossible
to say. No doubt there was some. It was attacked in the pages of the
Builder from time to time and pilloried in *Punch*. Legal proceedings
against it were occasionally reported in the press. Builders could be
prosecuted under local by-laws.

For example, at Edmonton, a northern suburb of London, several
cases were dealt with by the magistrates in July 1880. Edmonton was
growing fast as a working class suburb. The houses going up were
cheap ones. The first case as reported in *The Building News*, concerned
a builder, W. Cole, who was prosecuted by the local board of health
for erecting four houses of which the walls were not properly bonded
with mortar or cement. Evidence showed that the houses were built
of very inferior materials, small pieces of brick and an excessive amount
of mortar being used. The defendant was given three weeks to strengthen
the walls by covering them with cement which the board's surveyor
would check. When the case came before the magistrates three weeks
later the work had not been finished. The board secretary pressed for a
conviction and quoted a recent report on enteric fever in Edmonton—
'the cottages recently erected in Edmonton are many of them examples
of the flimsiest style of jerry building. The mortar used in some cottages
mainly consisted of road scrapings—black fetid sludge'. Cole was
thereupon ordered to pay a fine of £2. 10s. and costs of each of the
four summonses against him. In a second case William Gimson, also a
builder had been charged with eight summonses, four of them similar
to those of Cole, and four that the walls of the houses 'did not rest on
solid ground, concrete, or other sound foundation, as required by
by-law 101'. The magistrates had fined him £2. 10s. plus costs on
each summons. But Gimson had not paid and so the Bench granted a

warrant of distress on his property, such as it was, to recover the amount. On the same day the Board of Health applied for several other summonses. One of them caused laughter in court when evidence was given that a house was so faulty that the occupier when poking his own fire would also poke that of his neighbour at the same time. The Bench granted all the summonses, and expressed its satisfaction that the local Board of Health was acting so energetically to protect standards of housing in the district.

Jerry-building could also occur higher up the social scale: two drawings from *Punch* (see p. 124) illustrate the trials of the middle class householder.

Building of quite a different kind and on a bigger scale also went on in the London suburbs during the Victorian years. Many new churches were built in them from the eighteen-forties onward. The Church of England recognised the needs of the new populous districts and the political wisdom of meeting them. The Church Building Society received grants of £1,500,000 from Parliament and raised another £4,500,000 by the eighteen-thirties. Church and chapel attendance reached its peak in the mid-Victorian years. Also, educaton in England was backward compared with most industrial countries so there was plenty of scope for school building, especially after 1870 when at last an Education Act was passed. Under this act locally elected school boards could build elementary schools and there was a drive to provide a place for every child.

One firm which took advantage of these opportunities was Dove Brothers of North London. It was started by John Dove, a master carpenter of Sunbury, Middlesex, in 1791, at the time when John Nash was building prisons in Wales (see Chapter 1). For forty years the firm was quite small, doing general repairs and maintenance work. Then in the eighteen-thirties it joined in the church building boom and built its first church, St. Stephens, Islington, in 1838. From then on it specialised in this work and grew into a large firm. Forty years later it was building new churches at the rate of three or four a year, throughout the suburbs of London, as well as new schools and carrying out alterations and repairs to existing churches. The activities of the firm in the eighteen-seventies may be judged from the fact that it was submitting eighty tenders a year for large buildings, of which 10% were successful. In cash terms this meant work worth £36,000 per annum, at that time, each job being worth on average, £4,700. Skilful tendering and careful management ensured the firm's continual growth. When the boom in church building died away the firm had the special knowledge and skill to take a large amount of restoration work. (François Hennebique was building up his business on the same basis at about the same time in France—see Chapter 6.) Such work included St. Paul's Cathedral, Southwark Cathedral, St. Bartholomew-the-Great and St. Mary-le-Bow.

The same line of business continued in the twentieth century and up to the present day. In the early years of this century, the firm built the Central Hall and the Church House, both at Westminster, and more recently Guildford Cathedral, which was started in 1936.

THE BUILDING WORKERS AND THE TRADE UNION MOVEMENT (1859–1885)

1859 was a turning-point in the history of the building workers. For the first time in twenty-five years the craftsmen in all trades and the labourers united in one struggle. This was the lock out by all the large builders in London which lasted from July 1859 until February 1860. After this conflict the trade unions took a new path towards a more modern type of organisation.

The dispute occurred because of a rising demand for the nine hour day or fifty-four hour week. The usual hours of work were from 6.00 a.m. to 5.30 p.m. with $1\frac{1}{2}$ hours for meals. As we have seen in Chapter 2 there had been earlier attempts to win the nine hour day. The next best thing was to get the 'short Saturday' and, as a result of several small strikes in 1855 the masons, though not other trades, in London knocked off at 4 o'clock. However, in Manchester the next year a combined drive by all other trades won the Saturday half day. The London masons therefore petitioned the employers for Saturday work to finish at midday, but without success.

First to make a demand for the nine hour day in London were the carpenters and joiners through a committee of their clubs scattered throughout the city. This was refused. The carpenters therefore formed a joint committee with the bricklayers, and masons, and later plasterers, painters and even some labourers, and continued their agitation. The spark came from the employers who had become alarmed and anxious for a showdown. The joint committee petitioned four large employers that the hours of labour should be reduced by one hour per day:

'Gentlemen, we, the men in your employ consider that the time has arrived when some alteration in the hours of labour is necessary; and having determined that the reduction of the present working day to nine hours at the present rate of wages, asked for by the building trades during a public agitation of eighteen months, would meet our present requirements, we respectfully solicit you to consider nine hours as a day's work. A definite answer to our request is solicited by the 22nd July 1859.'

One of them, George Trollope in Pimlico, dismissed one of the masons who had presented the petition. When all the masons at Trollope's struck in support of their member, all the other trades followed.

The London Master Builders' Association, of London, held a meeting and resolved on a lock out, while at the same time employing non-union men. Their decision was warmly supported by Sir Morgan Peto, M.P. (late of Grissell & Peto). He spoke of the union's 'despotism'. He said that the time of the great mason's strike at the Houses of Parliament he had sacrificed £15,000 to £16,000 to fight them but he had never regretted it.

All the builders with more than fifty employees closed down and 24,000 workers were out of work. They also counter-attacked by presenting 'The Document' which every man would have to sign before returning to work, as twenty-five years earlier. This Document, produced by the Central Master Builders' Association, was printed in the form of books of tear-off sheets, with counterfoils which read:

'I declare that I AM NOT now, nor will I during the continuance of my engagement with you, become a MEMBER OF OR SUPPORT ANY SOCIETY which directly or indirectly interferes with the arrangement of this or any other Establishment OR the HOURS OR TERMS OF LABOUR, and that I recognise the rights of Employeers and Employed individually TO MAKE ANY TRADE ENGAGEMENTS ON WHICH THEY MAY CHOOSE TO AGREE.'

The dispute was therefore about the nine hour day and 'The Document'. No workers would sign the Document and a large part of the press and public opinion supported them. The employers therefore put a verbal declaration in place of the printed Document but this had no effect. They were intent on breaking the trade unions and so the workers concentrated on fighting the Document and dropped the nine hour day which many thought was not practicable.

They had to fight hard against the law as well as the employers. In November William Perham, secretary of the masons' union was convicted of intimidation at Clerkenwell Court and sent to prison for two months. The case against him was as follows. A certain Charles Robjohn had recruited eight men in Devon to work for Messrs. Piper, a builder in Bishopsgate, London, with a contract of six months' work at 5s. to 5s. 6d. a day. They arrived at Paddington Station at 9.30 p.m. on Saturday October 1st. There a member of the strike committee was waiting and he followed them in a cab to the Swan and Horseshoe in Little Britain where they were given supper. At this point Perham, who had worked for Pipers', appeared and said: 'He (Robjohn) has come to turn us out. You know me, my name is Perham. I have been out of work these eight weeks, and if you men go to work we shall call you "blacks", and when we go to work we shall strike against you.' This was enough. The result was that when Robjohn went to collect his men on the following Monday only four were still there, the rest had returned to Devon. Perham appealed against the conviction but lost.

The London unions would never have held out so long if it had not been for the money sent by committees all over the country; Glasgow raised £257, Blackburn £271, Manchester £545. Large sums also came from unions outside the building industry, for they realised that the Document was a basic threat. The London Society of Compositors sent £620, the Pianoforte Makers £300, the Shipwrights £300, but above all the Amalgamated Society of Engineers £1,000 a week for three weeks. Such a large sum from the engineering workers was a sensation and made a great impact on both sides in the dispute. Finally after seven months, in February 1860, the employers gave in and withdrew the Document. It was a drawn battle. That so much was won was largely due to the leadership of George Potter.

Potter was one of the big men who came from the building unions in this period. Born in 1832 at Kenilworth he was the son of a carpenter, was apprenticed to that trade and worked in it. He had some elementary education and could speak and write well. When he was twenty-one he went to London and soon became secretary of a small local club called the Progressive Society of Carpenters and Joiners. As such he became the very active and skilful leader of the London craftsmen during the dispute.

Immediately after it, unwilling to let go of the nine hour day, Potter organised a United Kingdom Association for Shortening the Hours of Labour in the Building Trades, trying to make it into a national movement. But the employers in London hit back by announcing that in future wages would be paid by the hour, instead of by the week. This was serious for it meant that not only the 'short Saturday' would be paid by the hour instead of as a full day, but that the workers could be dismissed at the end of an hour, particularly in wet weather. The bricklayers issued a notice:

'If we have no recognised number of hours a day, how can we withstand their capricious arrangements? The injury that can be inflicted upon us in winter we know to be immense.'

There were other strikes but they failed and the hourly system was brought in.

During the next twenty-five years the efforts of the building workers to win a better life centred round the following issues: piece-work, the new kinds of union, their part in the wider trade union movement, the legal position of unions, the franchise, the nine hour day. These efforts slowed down in the last ten years of the period because of the depression in trade.

Piece-work had been resisted by the workers since the early days of the industrial revolution, As we have seen, the stonemasons struggled against it. The battle was now continued by the carpenters and bricklayers. They were able to prevent piece-work being brought in and thus reduced speeding up, thanks to their increased strength. This resulted from the new kind of union which they formed. This in turn, was a result of the struggle in 1859–1860. During those months only the stonemasons had a national union; the other trades found how weak they were in small scattered clubs. Immediately after the strike the carpenters formed the Amalgamated Society of Carpenters and Joiners, a national body, and the painters, plasterers and bricklayers did the same. These new stronger unions, then started, lasted until 1921. It was a good time to start. The wealth of Victorian capitalism was increasing fast; there was more to be won if a cautious policy was followed; the old dream of a new society had disappeared. How was this new type of 'amalgamated' union different from the older ones? High contributions which only steady craftsmen could afford were the rule; as also were high benefits for sickness, unemployment, old age and death and these they would not want to lose. This resulted in caution, avoiding strikes, and a policy of industrial peace. The policy was also based on protecting the craftsmen against the unskilled worker. The A.S.C.J. is an example. This union was moulded by Robert Applegarth, its general secretary for nine years.

ROBERT APPLEGARTH (1834–1924)

Applegarth was the outstanding leader of the building workers in the mid-Victorian years and also well known in the wider world of the labour movement. He was born at Hull in the same year that the great Operative Builders' Union collapsed and he died two years before the general strike of 1926. His father was a seaman who sailed in the Greenland whalers and became captain of a tug. From 1839 to 1843 he was quartermaster on H.M.S. *Terror* in Ross's Arctic expedition in search of Sir John Franklin who had been lost with his crew in exploring the North-west Passage. Applegarth, as a boy, saw little of his father. When he was ten he started work in a shoemaker's shop where he was paid 2s. 6d. a week for blacking boots. Then he moved up to a merchant's office at 5s. a week and finally to a joiner's shop, where he was the boy who fetched the beer. He was not apprenticed to the trade but he learnt it, and after four weeks was earning 10s. a week. When he was eighteen he left Hull and made a home with his mother in Sheffield. As a skilled man he now earned 20s. a week but worked 60 hours a week for it. By 1854 he was a married man on 24s. a week.

That year, still under twenty-one, he was caught up in the great wave of emigration to America in the eighteen-fifties. Hundreds of thousands went to seek higher wages and a better life than they had in Britain. He landed at New York in December 1854, intending his wife to follow. There he was lucky enough to meet a Sheffield man, a manufacturer of powder flasks, who employed him fitting the leather to the flasks. Moving on into Pennsylvania he worked for a chair-maker and learnt the skill of turning chair legs. After moving on again to Chicago he settled down, making window sashes in the depot of the Chicago and Burlington Railroad, at Galesburg, Illinois, about 150 miles north-west of the city. At different times he also worked as a station master and in the engineer's office. After three years he was earning $2\frac{1}{2}$ dollars a day. Now it was possible for his wife to come but her health was too weak for the rough journey and so, in 1857, he returned to Sheffield.

The three years in the United States had taught Applegarth a good deal. Two events influenced him particularly. At the college of Galesburg he met the students. He and his mates exchanged knowledge with them. They taught the students the elements of their trade and in return the students taught them their theoretical knowledge. The second event was a first-hand experience of slavery. He went down the Mississippi, crossing from the 'free territory' of Illinois into the slave trading state of Missouri, to St. Louis to see for himself a sale of slaves.

He had not only read *Uncle Tom's Cabin*, published a few years before, but had met Frederick Douglass, the escaped slave whose memoirs of slavery inspired the anti-slavery cause. The Civil War was only four years away. At the sale the wind scattered the dealers' documents. Applegarth picked up and kept three receipts for slaves. He wrote: 'The lessons of Harriet Beecher to we and others had evoked in me such detestation of slavery that I felt I had the right to annex these receipts.' One of the receipts which he brought home read:

'Received of J. J. Williams, agent, 930 dollars, being the full purchase price for one negro named Caroline, the right and title of which slave I warrant and defend against claims by all other persons whatsoever and I do not warrant her sound.'

When Applegarth arrived in Sheffield trade was bad. The financial crisis of 1857, spreading from America, forced the bank rate up to 10%. He tramped to Manchester and back for a job but he was unemployed for some months. As soon as he found work he joined the Sheffield branch of the General Union of Carpenters and Joiners and soon after became its secretary and president. The first thing he did was to move the branch meeting place from its usual public house to a reading room where drink was not available, against the opposition of those who called him a 'whipper snapper who couldn't drink a glass of beer if he tried'. Intent on improving himself he made full use of the Free Library and helped to found a co-operative society.

Applegarth's union, the General Union was a weak one with the power in the hands of the local branch instead of a central committee and when the new model, the Amalgamated Society of Carpenters and Joiners was formed in 1860 as a result of the struggle of the previous year, he led his own branch into it. He saw the lesson of that struggle as follows:

'The London lock out induced a number of that trade to hold an inquest on the system of "localism" and their verdict was "the thing won't do".'

When Applegarth joined with the A.S.C.J. it had 618 members in twenty branches and only two of these, at Devonport and Kidderminster, were outside London. He became its general secretary two years later; he was then twenty-eight. Nine years later when he resigned from the secretaryship the union membership had risen to over 10,000

partly at the expense of its rival. These figures show the growth while
Applegarth was secretary:

Year	Number of Members	Number of Branches	Balance in Hand			Average Balance per Head		
			£	s.	d.	£	s.	d.
1862	949	38	899	8	10		17	10¾
1865	5670	134	8320	13	7	1	9	4½
1870	10178	236	17,568	19	4	1	14	10½

The branches included one in the city of New York, opened in 1867,
ten years after Applegarth had returned to England.

Applegarth had given up his work as an outdoor foreman at 34s. a
week to become secretary of the A.S.C.J. at 33s. a week. He paid for
lessons to improve his handwriting. The union never paid him more
than 50s. It gave him 7s. 6d. a week towards the rent of 13s. a week
for his home in York Street, Lambeth, in return for the use of a room
for meetings. He set about his task which he described as:

'To teach working men the practical lesson of self reliance, to provide
during the term of prosperity for the hour of need is one of our great
objects, the highest duty of Trade Unionists is to teach "man's duty to
man".'

For this, sound finance based on high contributions was necessary.
Good benefits were equally important, as seen in the union's emblem:
'In the Upper Panels is shown on the left a representation of a Workman having met with an Accident, borne away by his comrades; and
the companion subject on the right, shows the Workman disabled by
the loss of limb, receiving the Hundred Pounds Benefit at the hands
of the Treasurer of the Branch. The Lower Panels respectively show as
companion subjects, the Superannuation Benefit and the Relief of the
Widow.'

His attitude to strikes showed the same careful and cautious policy.
They were to be avoided in favour of argument and negotiations.
Arbitration and, as in his own case, emigration, were the cure for
industrial conflict. His own view was reported in the press as follows:
'With regard to strikes he would tell them at once that he did not
approve of that way of doing business, except in cases of absolute
necessity and when every other means had been tried and failed to
accomplish the desired object. If they had any grievances they should

write to their employers, and if they refused to agree to their terms, or took no notice of their appeals, the best thing was not to strike but to lay their claims before the public; and the masters would then be compelled to state their objections, on which the public would pass their opinion, which they might consider as the verdict of a jury.'

A typical example of his practice was the Cardiff joiners who were on strike for seven weeks in 1867. The question was whether the increased wage of 6¼d. per hour should be paid immediately or in six months' time. Applegarth settled it by compromising for payment in three months' time. But although he wanted co-operation between the two sides in industry he was quite clear that their interests must conflict:

'My experience has taught me that combinations result in increases in wages and decreases of hours. No sentiment should be brought into the matter. The employees are like the employers in trying to get as much as they can for as little as possible.'

During his long experience of negotiating with employers he once met John Brown, the great Sheffield steel master, a formidable opponent.

'You're from the Trade Union,' said Brown curtly.
'Yes,' admitted Applegarth.
'Well, I shall cut you short', Brown began.
'Pray don't,' replied Applegarth, 'I'm only five feet two, and that's short enough.'

He was a wiry, dark man with an ample beard, a big voice and much determination.

It was while he was general secretary of the carpenters and joiners that conciliation in the industry developed strongly. For some years the unions had been increasingly successful in getting employers to agree to working rules in a particular area. These rules had governed craftsmen's pay, limitation of apprentices and prevention of the use of half-finished parts. The employers had often resisted. Many said 'the code of rules proposed by the operatives is such an interference with the rights of employers and employed that they decline to accept them', and counter attacked with 'the Document'. However, the practice spread. Then in 1864 when there was a threat of a strike of carpenters at Wolverhampton a committee was set up, following action by the local mayor, composed of six members of each side to investigate the dispute. Both sides chose as independent chairman

Rupert Kettle, a judge at the Worcestershire county court. As a result the Wolverhampton conciliation board came into being and soon included also plasterers, bricklayers and masons. It succeeded in preventing disputes for ten years, when it broke down because the workers could not accept an adverse award and were dissatisfied with the long period of three years for which the awards were made. Its success for a time was largely due to Kettle's policy of getting agreement between the two sides and of using his power of arbitration as seldom as possible. This board was the start of widespread conciliation throughout the industry and gave a lead to other industries. Kettle received a knighthood for his work. In Parliament some attempts were made to encourage conciliation by the Councils of Conciliation Act, 1867, and the Arbitration (Masters and Workmen) Act, 1872, but in that age of *liassez faire* they had little effect.

While Applegarth and his fellow leaders of the builders' unions cautiously built up their strength they were at the same time members of the International Working Men's Association, the First International, which became a revolutionary body. The International, with its slogan of 'Workers of all lands, unite!' was started in 1864 by British trade unionists; Applegarth's own membership card was dated 1st January 1865, and he was chairman of its general council, when it met in Brussels three years later. Though Karl Marx in London became the leader of the International there was nothing revolutionary about it at first. Applegarth supported its demand for nationalisation of the land, and always stressed the workers' interest in peace and that they should oppose foreign policies leading to war. When the International grew to include many European countries and became revolutionary in its aims and tactics Applegarth still supported its general aims of the conquest of political power by the workers, long after his colleagues had resigned. Only after the Paris Commune of 1871 did he also leave the International though he had been attacked in *The Times* for his association with foreign revolutionaries. His object was to spread trade unionism and political democracy to the Continent. 'We in England', he told the International in conference at Basel in 1869, 'have no need to creep into holes and corners lest a policeman should see us'. Later he wrote:

'The great work of the International Working Men's Association lay in spreading the knowledge of the fact that the organised power of the workers was irresistible and that, if the workers would devote themselves to understanding the direction in which they should use that power, and unite in using it, they would soon make short work of the injustices of which they complain.'

As he was abroad in 1869, the *Sheffield Independent*, the *Scotsman* and the *New York World* employed him as their correspondent for the Franco-Prussian War which started in 1870. What he saw made him write:

'It is to my mind of the highest importance that, above all others, the working men of all countries should clearly and fully understand the miseries and hardships inflicted by war on themselves as a class.'

One incident particularly impressed him. The train on which he was travelling to Metz carried many German soldiers. When they got out to stretch their legs at Cologne Applegarth spoke to them:

'I could not resist the temptation to enquire the views of several on the general aspect of the war and what compensation they expected for themselves, and for their Country, for the tremendous sacrifices they were compelled to make. Several, who had no direct connection with the workshops, seemed rather to enjoy the attitude that they were going to "whip the Frenchmen", but others who had been called from the anvil, the bench and the loom expressed their regret that circumstances should compel the working men of two great nations to meet as foes. One of these men, a bookbinder at Leipzig, did not hesitate to denounce the war in the shortest terms. He wished to see the Imperial Government crushed and make way for a Government which, in his view, would be used to the interests of the working class than all the territory that could be taken from France. In saying goodbye to the bookbinder I whispered "You are a member of the International?" He said "Yes, are you?" I nodded and we parted with a hearty grip.'

Later when Applegarth was in Paris his passport was used to get members of the Government out of the city.

In the meantime, in England, Applegarth had been playing a leading role in other conflicts which were to him, along with most other Englishmen, much more important. These were about giving the vote to working men, and the legal position of trade unions. Applegarth had always believed that trade unions should always be concerned with politics. He said:

'I have wondered in the past when hearing members of Trade Societies cry out—No Politics—if they really understood what they said. . . . What is more natural than that trade unions should protect their members from bad laws as well as protect their wages?'

Thus in the same year that he became general secretary of the A.S.C.J. he and his friends had formed a political union to campaign for universal suffrage and vote by ballot. It was also natural when in 1865 the Reform League was established with a barrister as president and a bricklayer, George Howell, as secretary, to agitate for the vote, that Applegarth should be a foundation member and that the trade unions should be its main supporters.

The climax of the agitation was the demonstration in Hyde Park on 23rd July 1866, immediately after the conservative government had come into power. Afraid of rioting in the park, the Home Secretary ordered the park to be closed and the gates locked. The Reform League insisted on the right to hold its meeting. The demonstrators made their way to the park. Among them the banner of the carpenters and joiners on which Applegarth had had painted 'Deal with us on the square, you have chiselled us long enough', on one side, and on the other side 'Those who say politics will injure the Trade Unions have yet something to learn!' He was in the first carriage which had to stop at the locked gates. He got out and, engulfed in a crowd of a hundred thousand, was pressed against the park railings. He folded his arms to prevent his ribs being broken. The pressure was irresistible, the railings fell and were knocked down over a length of 1,400 yards, the crowd poured into the park and occupied it. This demonstration convinced the conservatives that reform was unavoidable and the Reform Act of 1867 gave the vote to the working men in the towns.

At the same time as these events the legal position of the trade unions came to a crisis. This was the turning-point in the history of trade unions. Applegarth had had plenty of experience of the law. 'At Sheffield', he said, 'I have known working men to be arrested in bed for absenting themselves from work and tried and sentenced in a magistrate's parlour before their families knew the offence with which they were charged.' This was the law of Master and Servant under which an employee could be prosecuted under the criminal law for breach of contract of employment. It was at Sheffield, too, in 1866, that the so-called 'outrages' occurred in which trade unionists used violence against blacklegs. This caused an outcry against trade unions. The next year there was another blow to the unions when the judges ruled that a trade union could not prosecute one of its officials who had made off with its money.

This reversed the previous position in which unions could protect their funds in this way. Applegarth issued a call to action:

'Let us then unite with dignified firmness and rest not until our unions have that protection to which they are entitled, and I trust that, with such protection and a few years' more experience, we shall have established a new era in the history of labour, have gained the full confidence of our employers, adopted arbitration as the first resort in our differences, and freed our unions from the expense and anxiety of strikes so far as it is possible to do so and we might then turn our attention to the establishment of a system that would embrace education for the young, employment for our surplus labour, the erection of meeting houses apart from public houses, as well as homes for our aged members.'

As a result, a royal commission was appointed 'to enquire and report on the organisation and rules of trade and other associations, with power to investigate any recent acts of intimidation, outrage, or wrong alleged to have been prompted, encouraged, or connived at by Trade Unions or other associations'.

Applegarth asked a member of the Government, Bruce, for the commission to include two trade unionists.

Bruce objected: They would be enquiring into themselves.
Applegarth: And very properly. They could give advice and asistance that could not be got from others. But my persuasiveness evidently will not get over your class prejudice. You would put employers on the commission, I suppose?
Bruce: Yes.
Applegarth: To condemn the class out of which they live?
Bruce: Capital employs labour.
Applegarth: And labour makes capital profitable.

The commission included only two sympathisers of the trade unions, one of whom was Thomas Hughes, M.P., Christian Socialist, author of *Tom Brown's Schooldays*, and later county court judge. But Applegarth and other trade union leaders gave evidence before it skilfully and gave a picture of unions as peaceful, law abiding, stable bodies.

One of the important issues was the right of picketing. On this the following exchange took place:

Applegarth: I say that it is perfectly justifiable for men to approach other men, to wait for them at the shop door and say to those who come—'We men were dissatisfied with the terms on which they were working at that place, and, if you go in, you will go in and undersell us; now we beg that you will not do that?'
Lord Elcho: 'Do you consider it legitimate picketing for men to assemble at the door of a shop where there are non-unionists and when they come out to howl at them?'
Applegarth: 'No. Speaking generally, I should say that howling would mean something offensive in itself.'

The majority of the commission were influenced by the evidence given by the Master Builders' Central Association. They recommended that the law should recognise trade unions and give protection to their funds but only if they did not control the number of apprentices, prevent the use of machinery, stop men doing piece-work, or maintain the closed shop. The minority on the other hand recommended that there should be no such restrictions.

Applegarth attacked the majority's report:

'If Parliament should agree with the majority that there should be one law for the rich and another for the poor, that Trade Unionists should be legislated for as a "dangerous class" instead of as citizens of a free country, and should enact a law . . . which would inevitably set one class against another in a spirit of determined hostility . . .'

then he would wish for its repeal and meanwhile the law would be evaded. In fact, a few years later, the Trade Union Act of 1871 and the Conspiracy and Protection of Property Act of 1875 gave the unions security for their funds and also made peaceful picketing legal so that the danger of prosecution for intimidation and violence was less.

Applegarth believed passionately in education and worked hard for the working class to have a share in it. In the events leading up to England's first Education Act in 1870 he played his part. In stating what the working class wanted—a national, free, compulsory, unsectarian system—he knew what he was talking about:

'The working classes have long declared in favour of compulsory education, and I should be sorry to be regarded as speaking in the interests of those I know little about, but my claim in speaking for the working classes is that I have worked with them and for them all the days of my life. . . .

'It is said that the abolition of school pence will disgrace us. I should like, gentlemen, to make that statement to a shop full of American workers who have had free education for many years. My experience is not that the American workers are degraded by free education but that many other workers are degraded by the lack of education.'

The Education Act was only a start, and a start made later than in the United States of Prussia. It was national and unsectarian, but it was neither compulsory nor free—school fees, the 'school pence', still had to be paid, unless the parents were too poor. The system was limited to children under thirteen. Applegarth's comment was prophetic:

'I regard the Elementary Education Act 1870 as but the beginning of a great national system: the children of the poor must be placed on a footing of equality with those of the rich, in having access to educational institutions for which their capacities may fit them from our National Infant Schools to what ought to be our National Universities.'

Before this he had already arranged for his own union to run classes for members in geometry, machine drawing and building construction. This was one of the first beginnings of technical education.

In the same year that as the Trade Union Act was passed Applegarth resigned from the secretaryship of his union. His task of establishing it firmly was completed for the time being. The circumstances in which he felt obliged to leave were, however, unpleasant. He was appointed a member of a Royal Commission, the first member of the working class ever to be so honoured. The Commission was to enquire into the law on Contagious Diseases, i.e. venereal disease. Applegarth's enemies in the union gained control of the executive and ordered him off the Commission because he would be wasting the union's time in investigating what they thought were obscenities. He refused and immediately resigned from his union post.

Applegarth continued to help workers in other industries, though he started a new career as an employer with his own small firm. In 1872 when the Gas, Light and Coke Company prosecuted the stokers on strike at Beckton Gasworks, London, for breach of contract and conspiracy, six of the leaders were sentenced to twelve months in prison. Applegarth collected money and got six others out of the country to Canada, New Zealand and Australia. In the same year he took a hand when Joseph Arch formed the first union for agricultural labourers. He spoke on the benefits of trade unions for a meeting in Norfolk. The vicar, who was in the chair, complained that Applegarth had not advised the men to save their money instead of spending it at the public house. Applegarth retorted:

'You have been told that you go to the public house and that I have not advised you to put money by for a rainy day. Men! if I were a labourer struggling to make ends meet on 13s. a week and I thought that, by going to the public house, I could purchase forgetfulness of my misery then by God, I'd go and I'd spend the lot.'

When the stonemasons at the new Law Courts in the Strand were on strike five years later against the contractors importing provincial men at a lower rate of wages, he intervened with Benjamin Hannen, secretary of the London Master Builders.

When he gave up being a labour leader Applegarth set the record straight on what it meant to be one:

'To those who think that the workers are led by the nose by their leaders, and that the life of the agitator is all "beer and skittles", I would say that if ever it falls to their lot to try their hand at leading and to test the quality of the "beer and skittles" they will find that it is one of the many ignorant fallacies that befog the public mind in connection with the labour question.'

He was only thirty-seven and had to make a living. He would have found it difficult to get employment so he set up as English agent for a Paris firm which produced breathing apparatus for poisonous atmospheres. It was widely used in England in mining accidents and also for underwater diving. He felt he was helping to improve safety in industry. Later he acquired the English patent of an early form of electric lighting. This was a kind of arc-light known as Jablochkoff's candle.

It was used at Billingsgate Fish Market and along the Thames Embankment and Holborn Viaduct. Applegarth carried out many installations, including the machine room at *The Times*. As a result he was elected a member of the Institution of Electrical Engineers.

Successful enough, by the age of fifty-six, he could afford to live in the country, at Epsom. He taught himself to ride. He refused to pay a fee of a guinea for the privilege of riding on the downs to the Grand Stand Association and successfully fought a local election on this issue. Hankering after the country life, he moved to Bexley, Kent, where on half an acre of land he kept poultry, even introducing a new type of hen from France. Finally he settled at Thornton Heath, Surrey, and there, when he was seventy-three, sold his business and enjoyed a long retirement of another seventeen years. In *Who's Who* he gave his recreation as: 'Work, more work, and still again more.'

As a result of his work the trade unions as they then were, i.e. unions of skilled craftsmen, took their place in the social system. He said:

'I have had some years of experience, and so long as the present re-lations between Capital and Labour continue, so long will Trade Unions be a necessity. The rising generation must be educated so as to make it fit to live in a higher state of social organisation. If you educate the workmen up to cooperative production, then the necessity for Trade Unions might cease to exist.'

While Applegarth was a union secretary the building workers played a bigger part in leading the general trade union movement of England. The strike of 1860 resulted in the London Trades Council being formed. 'At the termination of that memorable struggle', the Council stated, 'it was felt that something should be done to establish a general trades committee so as to be able on emergency to call the trades together with despatch for the purpose of rendering each other advice or assistance as the circumstances required.' The first secretary was a bricklayer, George Howell, who later became one of the first working men members of Parliament. At the same time there formed in London a group of trade union secretaries with their head offices there who came to dominate the whole movement. It was known as 'the clique', or by its enemies as 'the Dirty Pack' and later as 'the Junta'. The Junta controlled the London Trades Council which most trade unions had to ask for help in a strike. In it, together with the engineering workers and the ironfounders, were Applegarth, the carpenter, George Howell, the bricklayer, and also Edwin Coulson, another bricklayer and secretary of the Operative Bricklayers' Society. Coulson, described as 'bricky and stodgy' fell in line with the cautious policy of the Junta, but only up to a certain point. 'Capitalists tell us', he wrote, 'that the true interests of the workman lie in saving money, in using every effort to desert their own class and become masters. For what purpose? What does it profit us that half a dozen of our fellows in a generation should succeed in joining the war against the men who were formerly their comrades, and end, perhaps, by failing for half-a-million? No, we have a nobler morality and a higher aim than this: a feeling of brotherhood is the principle on which we will act, and our end shall be the elevation of our fellows—not into another class, but in their actions, their thoughts and their feelings.'

In the meantime, a fresh struggle for the nine hour day began. Trade and industry improved in 1869 and prospered until 1874. In 1870 there were several local strikes in the building industry but then the engineering workers started their successful strike for the nine hour day. This stimulated the building workers to join the general movement for shorter hours.

In January 1872 the London masons and carpenters and joiners demanded 'nine and nine', ninepence an hour and nine hours a day. They were joined by all the trades, except the painters who were too weak, and the employers locked them all out, about 5,000 men. But the men could not hold together. The masons, still thinking they were the aristocrats, betrayed the cause by negotiating separately and in July accepted 8½d. an hour. 'We think it almost unnecessary', they said in superior tones, 'to refer to the hubbub made by the other trades in London. We came to terms without consulting them.' The bricklayers and carpenters held out until August 31st but they too accepted 8½d. and hours which came to a little more than nine. The labourers gained a ½d. rise to 5½d. an hour and as a result tried to form a union on the same lines as the carpenters', but they were too poor and insecure for it to last.

The labourers of England remained poverty stricken and unorganised until the upsurge of the unskilled workers seventeen years later on. The craftsmen too remained quiet. From 1874 the country entered a long trade depression in which any long conflict was hopeless. An example was the stonemasons' strike from August 1877 to April 1878 on the building of the new Law Courts in Fleet Street for ½d. an hour more. It was a failure which cost the union £24,000. The union never fully recovered; its membership fell from 27,188 to 11,066 in six years.

By the end of this period, 1885, the position of the unions was roughly as follows. Many new members had flowed in during the good years but as many had left in the following years of depression. Membership figures of the main building unions in 1885 were:

Name of Union	Started in	Membership
Operative Stonemasons Friendly Society	1832	11,285
Operative Bricklayers Society	1848	6,412
General Union of Operative Carpenters & Joiners	1827	1,734
Amalgamated Society of Carpenters & Joiners	1860	25,781
Associated Carpenters & Joiners	1861	4,535
National Association of Operative Plasterers	1862	2,110
United Operative Plumbers Association	1832	2,666
United Operative Bricklayers' Trade, Accident, Sick and Burial Society	1832	1,975
		56,498

In building and construction as a whole about 875,000 men were employed and this number was increasing, but included a vast number of labourers who were not organised at all.

The total of 56,498 was about half of the membership of all unions in the building industry. The painters, for example, are not included above. They had always had great difficulty in organising themselves. A painter wrote in 1875:

'I believe the cause of the deplorable state of our trade is the amount of jerrying which is practised. It is no uncommon thing, on jobs which are contracted for by our employers, to have three and four coats of paint, not only to get two, but in many cases absolutely none. I need not here take up space in describing the very many ways in which this system of robbery is carried on, such as using size in the place of paint, whitening ceilings without first washing them off, and a variety of other ways which I am sure are familiar to the whole of our members. Hence it is that jobs that should last six months last three, and jobs that should last three months last six weeks. It is thus that our trade is so diminished that it will not allow those who follow it the bare necessaries of life.'

Because unions were almost non-existent labourers were employed to slap on the paint and painters' wages were lower than those of the other crafts. Some painters employed by large firms like Cubitts remained craftsmen but the work was increasingly done by labourers. Very slowly they formed small unions in London and Manchester and finally the National Amalgamated Society of Operative House Painters and Decorators which survived.

Two points stand out from the above table. The leadership had changed hands. The once proud and independent stonemasons had given way to the carpenters and joiners who were not only more numerous but better organised and financed. The bricklayers also were creeping up on the masons and would soon overtake them. This was because stonemasonry became less important than brickwork. Also, advance in some trades was held up by competition between unions in the same trade, e.g. carpenters. This problem was not to be solved for many years.

FURTHER READING:

H. J. Dyos, *Victorian Suburbs*, Leicester University Press 1961.
Punch LIX (21 Oct. 1871), 172.
Punch LXX (19 Feb. 1876), 64.
Asa Briggs, *Victorian People* (Chap. 7, Robert Applegarth and the Trade Unions), Pelican 1967.
G. D. H. Cole, *A Short History of the British Working Class Movement* (Vol. II), Allen & Unwin 1937.
T. J. Connelly, *The Woodworkers, 1860–1960*, Amalgamated Society of Woodworkers 1960.
A. W. Humphrey, *Robert Applegarth*, National Labour Press, 1914.
R. W. Postgate, The Builder's History—National Federation of Building Trade Operatives 1923.

5   Building with Steel

This chapter is about the U.S.A., and particularly Chicago, because it was there that the tall building, the skyscraper, was first erected, and on a steel frame. That was in 1885. It was only after twenty years that the first steel-frame building in Britain appeared, the Ritz Hotel in London. It was followed by many steel frame buildings in London and else-where.

How did this new method of construction come about, and in Chicago? Two men were particularly responsible, William le Baron Jenney and Louis Henry Sullivan, but the change would not have happened without the circumstances in which they worked. Those circumstances were the rapid onward rush of American industry and agriculture, at the same time as the frontier was being pushed ever westward, in the years after the end of the Civil War in 1865.

In Britain skilled labour was plentiful. In the United States it was scarce, and therefore there was a continual search for ways and means of doing without it. Technology made many skilled crafts obsolete. Because of this shortage of skilled labour in the vast spaces of America a typically American method of house building appeared early in the nineteenth century.

This was the light frame, or balloon frame as it was called in America. It was invented about 1835 by George Washington Snow, a surveyor and builder of Chicago and was known as Chicago construction. Whereas for centuries the house frame had been of heavy timbers joined by mortise and tenon joints, Snow's light frame was made of thin sticks and boards held together by nails. The skilled carpenter was replaced by the unskilled labourer and the cost was much reduced. The light frame made it possible for Chicago to develop very rapidly from a small village to a great city and for settler's houses to be built along the prairies. By the end of the century, however, Chicago con-struction meant a steel frame skyscraper.

WILLIAM LE BARON JENNEY (1832–1907)

Jenney, born at Fairhaven, Massachusetts, was a New Englander by birth and by descent. The Jenney family had landed at Plymouth, Massachusetts, in 1623. His second name came from a French surgeon who was shipwrecked on that coast and married into the family. Members of the family had fought against the British in 1812, although New England was opposed to the war.

Little is known about his childhood and youth. He went to Phillips Academy, Andover, Massachusetts and then to the Lawrence Scientific School, Harvard. As a young man he went to Paris, entered the Ecole Centrale des Arts et des Manufactures and graduated from it, with honours, in 1856. It so happened that in the same year two bigger events occurred which were to have an important effect on Jenney. These were the Crimean War and Bessemer's discovery of his process of steel making. The one led to the other. It was the request by the French for stronger cannon that resulted in Bessemer making his discovery. Later, it was Bessemer steel that Jenney used in his first steel frame building.

On his return to America, Jenney worked for a short time as engineer for the Tehuantepec Railroad Company of New Orleans which was building a railway in the isthmus of that name in Mexico. American capital was already finding a way into Mexico after the U.S.A. had annexed Texas in the war of 1848. But in 1858 Jenney went back to France where he studied architecture and also worked as an enginner for an American company. By this time he had received the best technical education available.

The American Civil War was a turning-point in his life, as with so many others. As soon as it broke out, in April 1861 he returned home and enlisted in the Federal Army. He was then twenty-eight. He went through several campaigns, took part in a number of battles, and served throughout the war. He spent five years in the army, first as a private, finally with the rank of major.

In the first Union campaign against the Confederate West, Jenney was in Grant's army which drove from Cairo, Tennessee, to Corinth, Mississippi. On the way he was present at the Union victories at Fort Henry on the Tennessee River, Fort Donelson on the Cumberland, and the bitterly contested battle of Shiloh in April 1862. Immediately after this last battle Jenney was commissioned with the rank of captain and taken on to General Grant's staff for engineering duties. The frequent river crossings by the armies gave him plenty of scope for his professional skill.

The whole of Jenney's service was in the western theatre of war. After the inconclusive battles of 1862 in Virginia, Grant's army of the

Tennessee opened the campaign to capture Vickburg early in 1863. The town was important for its control of Mississippi. It surrendered in May while General Sherman held off the Confederate army coming to relieve it. Jenney had been transferred to Sherman's staff at the general's request. Sherman, the tough little general, taught his officers to camp without tents just as the soldiers had to, and set an example himself. But one stormy night he had to break his rule; Jenney had had a tent erected for him against his orders.

After Vicksburg Jenney took part in the action at Jackson, some fifty miles to the west, and in the days of confused fighting about 200 miles to the north in the Mississippi valley in September. On one such day Sherman and his staff were going by train from Memphis westward to Corinth. Jenney, who was sitting near Sherman, started to eat a cake which had been given him in Memphis. Suddenly the train stopped. Enemy cavalry had cut the line. The cake was forgotten. The train was reversed to the next station where Sherman's force made a stand until reinforcements arrived. When the action was over Jenney looked for his cake. A soldier, who was eating it, called out: 'Who had the cake?' 'I did,' said Jenney. 'Well', mumbled the soldier, 'I got this out of a dead rebel's hand under the fence there. It's real good.'

Two months later Grant began the great battle of Chattanooga in the Tennessee valley, a vital point for the Confederacy. In the capture of Missionary Ridge Sherman, with whom Jenney was still a staff officer, had been ordered by Grant to continue attacking so as to draw the enemy forces. Having received an order to attack again Sherman told Jenney to signal Grant 'The orders were that I should get as many as possible in front of me. God knows there are enough. They have been reinforcing all day.' Grant replied: 'Keep pounding.' The men in the Union main force kept running up the steep rocky slope until they drove the enemy off the Ridge.

In May 1864, six months after Chattanooga, Sherman began his advance south-west on Atlanta in Georgia, which he took in September. Jenney had been detached from Sherman's H.Q. and posted to Nashville, Tennessee. There he was in charge of the engineers' department and map making. He remained there in comparative quiet while Sherman marched his army through Georgia and emerged on the sea at Savannah in December 1864. However, when in that month the fighting came to Nashville, Jenney was placed in command of the pontoon trains essential to rapid movement in a terrain with many rivers. When the Union forces were pursuing the Confederates after the battle he unwillingly carried out an order which allowed the enemy to escape. He was ordered to take a pontoon train south west along the Murfreeseboro road along there were no river obstacles, whereas the order should have been to go south-east to the Duck river. On the way he was asked where he was going. 'God knows', he shouted, 'where pontoon trains can do least possible damage.' He had queried the order twice and the general confirmed it twice.

Four months later, on 9th April 1865, the war ended when Lee surrendered to Grant at Appomattox. Jenney had been promoted to major and was chief engineer of an army corps. He stayed on Sherman's staff until he resigned in May 1866. He was thirty-three. Now after five years he had to make a civilian career. His experience had made him into a man who 'disposed of matters easily in the manner of a war veteran who believed he knew what was what'. He married, eventually had two sons, and in 1868 set up as an architect in Chicago.

For some years the buildings Jenney designed were quite conventional small churches and office buildings. The very year after he started in practice he published jointly a handsome volume entitled *Principles and Practice of Architecture* with forty-six plates and plans of churches, dwellings and stores constructed by the authors. The chief exhibit, the Grace Episcopal Church, built of stone, was in an imitation French Gothic style of the thirteenth century. But he made progress. He introduced the use of pressed brick in office buildings. It was one of these that attracted the young Louis Sullivan when he arrived in Chicago in 1873 and led him to ask Jenney to employ him. At that time Jenney employed five men and a boy, including some who were to be famous architects. Sullivan retained a vivid impression of his employer:

'The Major was a free-and-easy cultural gentleman, but not an architect except by courtesy of terms. His true profession was that of engineer. He had received his technical training, or education, at the Ecole Polytechnique in France, and had served through the Civil War as Major of Engineers. He had been with Sherman on the march to the sea.

'He spoke French with an accent so atrocious that it jarred Louis's teeth, while his English speech jerked about as though it had St. Vitus's dance. He was monstrously pop-eyed, with hanging mobile features and sensuous lips. Louis soon found out that the Major was not, really, in his heart, an engineer at all, but by nature and in toto, a *bon vivant*, a gourmet. He lived at Riverside, a suburb, and Louis often smiled to see him carry home by their naked feet, with all plumage, a brace or two of choice wild ducks, or other game birds, or a rare and odorous cheese from abroad.

'And the Major knew his vintages, every one, and his sauces, every one; he also was master of the chafing dish and the charcoal grille. All in all the Major was effusive; a hale fellow well met, and an officer of the Loyal Legion, a welcome guest anywhere, but by preference a host. He was also an excellent raconteur, with a lively sense of humour and a certain piquancy of fancy that seemed Gallic.'

Jenney was, as Sullivan noticed, one of the few 'who were intelligently conscientious in the interest of their clients'. The new type of building which he constructed came about in response to the interests of his clients, merchants, financiers, brokers in the humming city of Chicago. The first signs of this was an eight storey warehouse he built at 280 West Monroe Street in 1879 which was still standing in 1959. The method of construction was not new; the exterior was supported by brick pillars, and granite piers on the ground floor, the interior by iron columns as was quite common. Only the unusually wide glass openings in the walls pointed to the future. The load was not entirely carried by an iron frame. About 18% of the weight of the building was carried by the ground floor piers. In his next building Jenney broke away from this; the whole weight was carried by an iron and steel frame.

This innovation, in the ten storey Home Insurance Company building, Chicago, was due to the company's requirements. The company wanted a new type of office building which would be fireproof and also have numerous small rooms each having the maximum light. This meant that Jenney had to reduce the thickness of the piers between the windows. This, in turn, meant he had to use cast-iron columns instead and so logically, a frame or cage to carry floors as well as walls. He surrounded the columns with brickwork which supported only itself and not the adjoining floors. The external walls were merely panels between the members of the frame. In this way, he evolved the new construction. It was built between 1883 and 1885.

In this building Jenney used a new material, steel, as well as a new method of construction. For the lower six floor he used old fashioned wrought-iron floor beams, but then the great Carnegie Works at Pittsburgh, where Andrew Carnegie was the super salesman always looking for a new market, asked if they could supply steel beams. Jenney agreed and they were used for the construction of the upper floors. This was the start of the steel skyscrapers. Of course the tall

buildings of this kind would not have been put up unless there were
safe passenger lifts. Since about 1850 there had been goods lifts which
worked on a hydraulic motor and a system of pulleys but this meant
that the cage hung at the end of a rope, which was not safe enough for
use by passengers. The necessary safety device was invented by Elisha
Graves Otis and installed by him in New York in 1857. It was an arrangement of ratchets on each side of the lift shaft, on which pawls
on the cage engaged when the rope failed.

After the success of this building Jenney's services were in great
demand in Chicago. He took into partnership W. A. Otis and in 1889
designed the Leiter Building which took the steel frame or skeleton
a stage further. (At about the same time the first electric lifts were
installed.) This was the first building in which there was not even one
self-supporting wall. It was reported in a contemporary journal:

'The building will be located on the East side of State Street and will
extend from Van Buren to Congress Street. It will be eight storeys;
the three street front will be of light grey New England granite, the
construction a steel skeleton, the masonry protecting the external
columns. It is intended that the whole building should be one great
retail store.'

The front, 400 feet long, made no attempt to disguise the steel frame.
The huge squares of the frame were, on the exterior, simply filled with
windows merely separated by metal columns. The building made
a big impression; a contemporary writer was enthusiastic:

'It has been constructed with the same science and all the careful
inspection that would be used in the construction of a steel railroad
bridge of the first order. The severely plain exterior is grand in its
proportions. The great corner piers are carried upward to a chaste
cornice. Designed for space, light, ventilation and security, the Leiter
Building meets the effect sought in every particular. . . . A giant
structure . . . healthy to look at, lightsome and airy whilst substantial;
was added to the great houses of a great city.'

Other Chicago buildings by Jenney followed fast. The Manhattan
was designed in May 1890 and completed in the summer of 1891.
It stood, 204 feet high, in Dearborn Street. As this street was rather
narrow it was necessary to catch as much sunlight as possible. Jenney
therefore departed from the perfectly flat wall surfaces of his previous

buildings which corresponded to the frame, and inserted bay windows. In his next building, however, in 1891, the Fair Building, he returned to his previous 'severely plain exterior'. In this nine storey building the two lowest storeys were almost entirely glass to meet the requirements of the client who wanted the maximum display space.

Jenney had provided a new method of construction and a style to go with it. Across the way from his Fair Building there had gone up at the same time the massive Monadnock Building, fifteen storeys high but built of masonry. The piers were fifteen feet thick at the bottom. It was the last of the high buildings with walls of solid masonry. Thereafter they were constructed on the steel frame pioneered by Jenney.

A few years later the Bessemer Steamship Company named a vessel after him in recognition of his services to the steel industry.

He continued a busy life until he retired to Los Angeles when he was seventy-three. Various other buildings were for banks and insurance companies. His last work which he could not finish because of ill health, was a link with his Civil War days—the Illinois Memorial on the battlefield of Vicksburg.

LOUIS HENRY SULLIVAN (1856–1924)

Sullivan's father was Irish, his mother Swiss. His father, Patrick Sullivan, after wandering in Europe playing his fiddle and teaching dancing, emigrated with the flood of Irish in the eighteen-forties. His mother came from a middle class family in Geneva who lost their money and emigrated to make a fresh start in the new world. She was a skilful pianist and together they set up a school of dancing in Boston, Massachusetts, where Louis Sullivan was born.

Because his parents were frequently on the move Sullivan spent most of his childhood with his grandparents on their farm ten miles from Boston. There he led a happy, free, country life. From time to time his parents took charge of him when, as often happened, they moved to open a new school. This meant that he attended one school after another. Here is his picture of his early village school:

'The school-room was large and bare with two wooden posts supporting the roof. The teacher sat at her desk on a raised platform at the wall opposite the entrance. The children sat at rows of desks (a row per grade) at right angles to the rear wall; in front of them an open space for recitation by class; blackboard on the wall and so forth.

There were five grades in the single room. Teacher sat at her desk, ruler in hand to rap with or admonish. All the children studied their lessons aloud, or mumbled them. The room vibrated with a ceaseless hum, within which individual voices could be heard. Everything was free and easy—discipline rare. There was however a certain order of procedure. Came time for a class to recite. They flocked to the wall and stood in a row; neither foot nor head at first. Questions and answers concerning the lesson of the day. Teacher's questions specific; pupils' answers must be definite, categorical. Teacher was mild, patient; the answers were sometimes intelligent, more often hesitant, bashful, dull or hopelessly stupid. Each answer was followed by a monotonous "Go to the foot", "Go to the head"; and all the time the hum went on, the unceasing murmur, a thin piping voice here, a deeper one there, a rasping out yonder, as they pored over their primers, first readers, geographies, arithmetics; while now and again Teacher's voice rose high, questioning the class on the rack, the children answering as best they could. This babel merged or deliquesced into a monotone; there seemed to be a diapason, resonant, thick, the conjoined utterance of many small souls trying to learn, entering the path of knowledge that would prove short for most of them. The children were all barefoot and rather carelessly clad; notably so in the matter of omissions.'

Later, at the grammar school in Boston he learned quickly, and at the age of fourteen he passed on to the High School. He had a good reason for wanting to make progress. He had already become interested in buildings and had seen an impressive figure of an architect coming from one. Conversation with a workman convinced him that he would be an architect and that he had to have more education first.

The master of the first year class at the High School, Boston, was an inspiring teacher and he had a lasting influence on Sullivan. The subjects of algebra, geometry, English literature, botany, mineralogy and French as he taught them interested Sullivan all his life. He did not finish the course at the High School, but when he was sixteen he entered the Massachusetts Institute of Technology, having easily passed the entrance examination.

At the 'Tech' Sullivan enrolled in the school of architecture, the first in America. Although he enjoyed the freedom there he found the staff inferior, the course uninteresting and the school, in his words, 'but a pale reflection of the Ecole des Beaux Arts'; a mere copy of what was taught in Paris. He was in a hurry, he therefore left the college after only one year and decided to get some practical experience. He had learnt to draw well.

Arriving in Philadelphia he noticed a particular house being built in a way which appealed to him and got the architect to employ him at 10 dollars a week. There he was thoroughly trained in draughtsmanship. But the job was not to last. In that year, 1873, one of America's frequent financial crises occurred. The boom after the Civil War which was based on the building of many thousand miles of railway, led to an orgy of speculation and collapse of the banks. Sullivan was out of a job. He took the Pennsylvania train to join his parents in Chicago.

That city was recovering from the Great Fire of 1871 which had destroyed its centre. It had been largely built of wood. Even where the materials were brick, timber and iron, the iron columns and beams had melted where the timber floors caught fire. Sullivan explored the ruins; he thought 'this is the place for me'. He wrote of:

'An intoxicating rawness: A sense of big things to be done. For "Big" was the word. "Biggest" was preferred, and the "biggest in the world" was the braggart phrase on every tongue. Chicago had had the biggest conflagration "in the world". It was the biggest grain and lumber market "in the world". It slaughtered more hogs than any city "in the world". It was the greatest railroad center, the greatest this, and the greatest that.'

Sullivan repeated his action in Philadelphia. Having seen a building he admired, he applied to the architect, William le Baron Jenney, and was immediately taken into his office where there were five men and a boy. Sullivan learnt a great deal there. He went to concerts and became a great lover of Wagner's music. But, still restless, he wanted to increase his knowledge of European theory which held sway in America. So after only a few months, in 1874, he sailed for England en route for Paris. He walked all over London and was astonished at:

'The surging crowds, the dismal hardness of so many faces, and a certain ruthlessness; and everywhere in the jammed highways, the selfish push of those who must live, the shoals of wretches in the Haymarket clutching at his sleeves, so he confined himself to the pleasanter aspects, such as Hyde Park, Rotten Row, and the Thames embankment. He was curious at the vast Houses of Parliament, vertical everywhere; and St. Paul's black with soot; and many structures in which he sensed, in their visages, the solemn weight of age. They did not appeal to him in their historic message so much as in the sense of that which is old. This massive oldness made a new sensation for him. So passed the days.'

In Paris Sullivan took a room on the seventh floor of a small hotel in the Latin Quarter. He was now only eighteen. After several weeks of special tuition in mathematics and French he passed the difficult examination for entry into the Ecole des Beaux Arts and became a pupil of a famous architect. He read widely in history and philosophy, visited Rome to see the work of Michaelangelo and Florence. But much as he enjoyed the free life of the studio, he soon realised that the teaching was not what he was looking for. He found brilliant technique but a lack of imagination. So again he left early, after only six months, and by March 1875 he was back in Chicago.

For four years he gained experience as a draughtsman in a number of architects' offices. Through friendship with an engineer, he became more interested in engineering than architecture and thought of being a bridge engineer. In his own words he 'found himself drifting towards the engineering point of view, or state of mind, as he began to discern that the engineers were the only men who could face a problem squarely; who knew a problem when they saw it'.

The problem with which Sullivan was trying to grapple in his mind was a big one. It was: what part would he play in the rebuilding and rapid growth of Chicago; and also, more important to him, how could he develop his own style, a style which would be right for the tall buildings to go up. He was searching for a new principle of design which he could apply to all these new structures which were of a different kind, in their height, size and materials, from what had gone before.

At the same time Sullivan was in a hurry to advance in his profession. In his autobiography he wrote:

'His plan was, in due time to select a middle-aged architect of standing and established practice, with the right sort of clientele; to enter such an office, and through his speed, alertness and quick ambitious wit, make himself so indispensable that partnership would naturally follow.'

His plan came off. He soon met an architect, Dankmar Adler, who after being like Jenney, an engineer in the Civil War, had built up a practice in Chicago. He went into his office and was so useful that in 1881, when he was still only twenty-five, he was taken into partnership and the firm of Adler and Sullivan was formed. He had always been precocious. Adler encouraged Sullivan to develop his abilities and the two worked well together.

The world lay at his feet. Sullivan had now worked out his principle of design. He wrote:

'The forms under his hand would grow naturally out of the needs and express them frankly, and freshly. This meant in his courageous mind that he would put to the test a formula he had evolved, through long contemplation of living things, namely that *form follows function*, which would mean, in practice, that architecture might again become a living art, if this formula were but adhered to.'

He meant to have a new style of building, which would be truly American, and freed from the imitation classical and Gothic influence from Europe.

The *function* that was becoming all important in Chicago was office administration. The population of the city grew from 4,000 in 1837 to half a million in 1880 and a million in 1890. It was a great centre for banking and insurance and transport as well as for grain, timber and meat packing. The number of cattle slaughtered by Chicago packers quadrupled between 1875 and 1890. The demand for high office blocks increased in step with this growth. The values of city sites soared. High buildings would increase the profits of owners of real estate, and investors, speculators and developers abounded. They were made technically possible by three things. The invention of hydraulic lifts; the light or skeleton framework, which Jenney pioneered; and the rapid increase in production of steel by the Bessemer process. After the Civil War the telephone had already arrived and electric light was spreading. Out of all these came the skyscraper in Chicago.

It was ten years before Sullivan could put his ideas into practice. At that time a number of brilliant engineers and architects were trying to find ways of building higher and higher. The buildings were still of the composite type in which the load was shared between a frame of iron columns and beams and masonry or brick piers. Increasingly the load was shifted to the frame and buildings were pushed up to twelve storeys. But this weight created problems for the foundations in Chicago's mud, and the thick masonry still required

took up too much space. Jenney solved this problem with his Home
Insurance Building on an iron and steel skeleton.

While Jenney's revolutionary construction was being built, Sullivan
and others were still designing buildings of the composite type—
a number of office blocks. Among them was the vast Auditorium
Building on which he and Adler laboured for four years (1886–1890).
Sullivan wrote:

'For several years there had been talk to the effect that Chicago needed
a grand opera house; but the several schemes advanced were too aristocratic and exclusive to meet with general approval. In 1885 there
appeared the man of the hour, Ferdinand W. Peck . . . he wished to give
birth to a great hall within which the multitude might gather for all
sorts of purposes including grand opera . . . the theatre seating 4,250
he called the Auditorium, and the entire structure comprising theatre,
hotel, office building and tower he named the Auditorium Building—
nobody knows just why—anyway it sounded better than 'Grand
Opera House'. . . . Came the Auditorium Building with its immense
mass of ten storeys, its tower, weighing thirty million pounds, equivalent to twenty storeys—a tower of solid masonry carried as a "floating"
foundation; a great raft 67 by 100 feet.'

This was one of the last buildings with load bearing walls, which
were heavy as well as high. Thereafter they were in a lighter, steel
construction. Adler and Sullivan were exhausted by the Auditorium.
Sullivan travelled to California, thence to Mississippi where he built
himself a cottage on the quiet backwaters of Biloxi Bay. Nearby was
Ocean Springs:

'. . . a village sleeping as it had slept for generations with untroubled
surface; a people soft-voiced, unconcerned, easy going, indolent;
the general store, the post office, the barber shop, the meat market,
on Main Street, sheltered by ancient live oaks; the saloon near the
depot, the one-man jail in the middle of the street back of the depot;
shell roads in the village, wagon trails leading away to the hummock
land; no "enterprise", no "progress", no booming for a "Greater
Ocean Springs", no factories, no anxious faces, no glare of the dollar
hunter, no land agents, no hustlers, no drummers, no white-staked
lonely subdivisions.'

Sullivan described the land he bought:

'What he saw was not merely woodland but a stately forest, of amazing beauty, utterly wild; immense rugged short-leaved pines, sheer eighty feet to their stiff gnarled crowns, graceful swamp pines, very tall, delicately plumed; slender vertical Loblolly pines in dense masses; patriarchal sweet gums and black gums with their younger broods; maples, hickories, myrtles; in the undergrowth, dogwoods, Halesias, sloe plums, buck-eyes and azaleas, all in a riot of bloom; a giant magnolia grandiflora near the front—all grouped and arranged as though by the hand of an unseen poet.'

The cottage was Sullivan's retreat for eighteen years until it was wrecked by a hurricane.

On returning to Chicago Sullivan found that the steel frame construction for tall buildings was now being used by one or two architects. The Bessemer process in the steel mills of Pennsylvania had been producing mild steel for some years but mainly as rails; structures had still been made of iron. Then the sales managers of the mills saw the possibilities of the high building in Chicago. They already sold steel members for bridges and it was an easy matter to offer them to architects and builders.

Sullivan felt that this new, revolutionary form of construction called for an equally revolutionary kind of design. The first signs were seen in his Wainwright Building in St. Louis, Missouri. This was a ten storey building containing shops on the ground floor and offices above erected in 1891. It was still a compromise in style between the traditional and the new. The load was borne by the steel skeleton but there were also tall red brick piers which seemed also to bear the weight. Sullivan used them to stress the loftiness and upward sweeping lines of the building. 'What is the chief characteristic of the tall office building?' he asked. 'And at once we answer, it is lofty. . . . It must be every inch a proud and soaring thing. . . .'

Two more tall buildings, designed by Sullivan, followed rapidly in the next three years, both in Chicago. One was a seventeen storey tower, the other, a twelve storey rectangular block, the Stock Exchange. These led to his best skyscraper, built in 1894–5. This was the Guaranty (later Prudential) Building in Buffalo, New York State, the great centre of milling where ten mills produced three-quarters of a million barrels of flour. This thirteen storey building was the nearest he got to his ideal of combining the practical functions, of a skyscraper with his theory of 'form follows function' and 'a proud and soaring thing'. The practical requirements he defined as follows. These he succeeded in giving their most satisfying and attractive clothing:

'First, a storey below ground, containing boilers, engines of various sorts, etc.—in short, the plant for power, heating, lighting, etc. Second, a ground floor, so-called devoted to stores, banks, or other establishments requiring large area, ample spacing, ample light, and great freedom of access. Third, a second storey readily accessible by stairways—the space usually in large subdivisions, with corresponding liberality in structural spacing and in expanse of glass and breadth of material openings. Fourth, above these an indefinite number of storeys of offices piled tier upon tier, one tier just like another—an office being similar to a cell in a honey-comb, merely a compartment, nothing more. Fifth and last, at the top of this pile is placed a space or a storey that, as related to the life and usefulness of the structure, is purely physiological in its nature—namely, the attic.'

The clothing which Sullivan gave in the Buffalo Building to these practical functions deliberately stressed, with its piers and sheathing in highly decorated terra cotta, its vertical lines, so as to give his soaring effect. Two years later the only example of his work in New York appeared; the twelve storey Bayard Building on Bleecker Street. This aimed at the same effect, the steel piers topped by elaborate ornament and tracery, like lofty trees bursting into plumes of foliage. At the same time, however, Sullivan came to realise that his stress on vertical lines in a building did not fit the basic structure of the steel frame. The frame was nothing more than a case in which horizontal lines fitted the structure as much as the vertical.

In his next building he achieved a balance between the two. This was the Schlesinger and Mayer department store on the 'World's Busiest Corner' at the junction of State and Madison Streets in Chicago, twelve storeys high, and extending along both. The function was display and sales. Inside, the large unobstructed areas on each floor, with plenty of daylight, were made possible by the steel frame of the building. Externally, the balance between horizontal and vertical lines was made by the wide horizontal 'Chicago windows', spaced so as to coincide with the steel frame and the narrow piers which separated them. Each window was surrounded with a thin frame to stress the honeycomb nature of the frame. The simple plainness of this exterior was offset by the two lowest floors which were faced with a screen of cast iron and glass, the cast iron being covered with fantastically elaborate ornamental designs in which Sullivan, as usual, gave full rein to his fancy and love of natural shapes.

This was his last big city building, although at that time, 1900, he was only forty-four and at the height of his powers. His career was already in decline. He had received more than a hundred commissions but from then on he received a mere twenty. The basic reason was that his style was too original and his personality too individualistic to compromise with the times he lived in. He was already in conflict with the trends of the turn of the century. This conflict had started with the Chicago World's Fair of 1893, called the World's Columbian Exposition which was to celebrate the anniversary of Christopher Columbus's discovery of America. The organisers decided that the exhibition buildings should be in the Roman classical style, in favour with the financiers of New York, in spite of the new American style which had appeared in Chicago and elsewhere. Sullivan was the only man to condemn this and to design his own Transportation Building in the contemporary style. He made enemies by his scathing criticism of the exhibition which was greatly admired by the crowds who flooded to it:

'. . . the structure representing the United States Government was of an incredible vulgarity, while the building at the peak of the north axis, stationed there as a symbol of 'The Great State of Illinois' matched it as a lewd exhibit of drooling imbecility and political debauchery... placed on the border of a small lake stood the Palace of Arts, the most vitriolic of them, the most impudently thievish, . . . No keynote, no dramatic setting forth of that deed (the voyage of Columbus) which, recently, has aroused some discussion as to whether the discovery of America has proven to be a blessing or a curse to the world of mankind.'

Sullivan recognised the forces against him. He wrote later:

'The damage wrought by the World's Fair will last for half a century from its date, if not longer . . . there came a violent outbreak of the Classic and the Renaissance in the East, which slowly spread westward, contaminating all that it touched, both at its source and outward. The selling campaign of the bogus antique was remarkably well managed through skilful publicity and propaganda by those who were first to see its commercial possibilities. Thus architecture died in the land of the free and the home of the brave—in a land declaring its fervid democracy, its inventiveness, its resourcefulness, its immense daring, enterprise and progress. Thus did the virus of a vulture, snobbish and alien to the land, perform its work of disintegration.'

And thus Sullivan fell out of favour. The growing fashion of imitating Classical and Gothic styles from Europe left him isolated. When he separated from Adler who supplied great technical skill, he did not form another partnership. This was a handicap in the age of bigger and bigger corporations and the growth of powerful trusts in industry. His personal life became unhappy. His marriage in 1899 was not a success; he drank heavily and his wife left him a few years later. His practice in Chicago disappeared, he lived in a cheap hotel and eventually he was helped with money by his colleagues.

However, he designed some more remarkable buildings. But these were all in the Middle West; they were small banks in little country towns. There, in rural surroundings, far from the influence of Chicago and the big cities he found clients more friendly towards his ideas and his ideal of a native American design freed from the trappings of the old European culture. The banks were farmers' banks. The towns were in the midst of the great corn belt which stretched from Western Ohio to South Dakota, including most of Indiana, Illinois, Iowa and southern Minnesota, as well as eastern Kansas, Nebraska and northern Missouri. There the products on a vast scale were Indian corn, or maize, and hogs. Sullivan felt freer in the more democratic surroundings.

The first bank built in 1907–1908 was the National Farmers' Bank at Owatonna, Minnesota, fifty miles south of Minneapolis. The building was a square block, in fact almost a cube. Above the base of ashlar sandstone rose the large walls of rough shale brick. The main effect came from the arched windows, two great semicircles, each spanning 30 feet and filling a great part of each wall. In each of these windows there were fourteen slender vertical steel mullions. These simple geometric lines made a sharp impression. They were repeated in the interior where Sullivan also indulged his love of decorative patterns of leaves and scrolls.

Sullivan followed much the same simple yet dramatic design, with many variations, in his other bank buildings. The main ones were: the People's Savings Bank at Cedar Rapids, Iowa, 1911; the Merchants' National Bank, Ginnell, Iowa, 1914; the People's Savings and Loan Association Bank, Ohio, 1917–1918; the Farmers' and Merchants' Union Bank, Columbus, Wisconsin, 1919.

The names of these banks suggest the grass roots of America, the ordinary people of the prairies. Sullivan believed that in some way his buildings were the right and appropriate ones for democratic America and therefore he was rejected by the financiers and capitalists of the big cities. In the long run his work led to that of a great designer of houses, Frank Lloyd Wright.

THE BUILDING WORKERS IN THE U.S.A. (1880–1900)

In spite of the changes in technology and materials during the years of Jenney, Sullivan and the Chicago school, the building workers kept their strong position. In fact they improved it. The craftsmen's unions increased their membership to a figure of over 390,000 by 1904. The construction men also formed their first union. During the eighteen-eighties and 'nineties there was bitter conflict between labour and capital. The rise of the American Federation of Labour provoked a strong anti-labour movement amongst business men. There was bloodshed, as among the workers in the Carnegie steel works at Homestead in 1892 and the Pullman workers near Chicago in 1894.

In the chaotic conditions of a rapidly growing building and construction industry the strength of the unions led to serious conflicts in Chicago and New York. The slump of 1893 and the armies of unemployed, one of which, led by an army general, marched on Washington, deepened the general unrest.

The bricklayers and the carpenters led the movement in the building industry. The bricklayers had started a national union, the Bricklayers' and Masons' International Union of America in 1865, immediately after the Civil War. The carpenters found it more difficult. One reason was that the Amalgamated Society of Carpenters and Joiners of England had branches in America. Gradually, however, it was overtaken by the native United Brotherhood of Carpenters and Joiners which had been founded in 1881. The other crafts, painters, plasterers, plumbers, were all able to set up national unions in the decades after the Civil War. At the same time the unions established local trades councils. One of the first was in Chicago, the Amalgamated Council of Building Trades of 1887. As well as the trades just mentioned the council included a wide range of occupations—stonecutters, derrickmen, hod carriers, lathers, gas fitters, galvanised iron and cornice workers, slaters, stairbuilders.

The employers were, like those in England, of two kinds, a few large firms, and a great many small ones. Another similarity was that contractors for the brick work usually undertook a building and sub-contracted the other work. This meant that the employers found it difficult to form associations, except on a local basis. However, at the same time as the formation of the Chicago trades council they did form a national association of builders in Chicago.

Conflict between the two sides broke out in 1887. A strike by the bricklayers for the eight hour day and against the employment of unskilled labour lasted eleven weeks. By arbitration the eight hour day was accepted. The difficult position of the foremen was also settled; he was to be a member of the union but not subject to the union rules while he was working for a contractor. This agreement lasted ten years.

Four years later, the trades council re-formed into a stronger body, the Building Trades Council; 'for the purpose of assisting each other when necessary, thereby removing all unjust and injurious competition, and to secure unity of action for their mutal protection and support'. The council prevented non-union men from working whenever it could. In Chicago, it became a powerful body by enforcing rules on its members. Two of these were that non-union men were to obey the rules of the union in their trade, and that no trade was to work under police protection or use materials made by convict labour. The council's power increased when the unions of men making builders' materials, brickmakers and timber workers, also formed a building materials trades council. It could then prevent employers from using materials made by non-union labour.

Conflict was bound to arise and it came when the employers organised their own council in Chicago in 1899 in order to attack the unions. The points at issue were basic. They were: restrictions on the use of machinery, limitations on the amount of work done and the number of apprentices to be employed, so-called interference by the unions with men at work, and sympathetic strikes. The employers issued a statement that they would not allow any of these.

Agreement was reached with the trades council but the unions affiliated to it refused to fall in line. The employers therefore repeated their statement as an ultimatum, adding to it the demand that they would have not only the complete right of hiring and firing but also of deciding what materials were used, whether made by union labour or not. On the other hand they agreed to an eight hour day, time and a half for overtime, and double time for Sundays and holidays. The unions refused, however, to accept this package. This resulted in a lock out which lasted ten months, from February 1900 to April 1901 and affected 50,000 to 60,000 workers. It was a bitter struggle. There was much violence, typical of American industry at that time; four people were killed and many injured. The result was defeat for the unions. The Chicago employers refused to negotiate with the trades council. First the bricklayers, then the other trades, withdrew from the council and had to accept the employers' demands.

In the meantime there was a pointer to the future in another direction. The new methods and new materials in tall buildings brought about a new organisation of labour. Very soon after the first steel frame buildings appeared in Chicago the Bridge and Construction Men's Union was founded in that city in 1891. These men who had worked on the construction of bridges, first of timber then of steel, began to work on buildings, in the same way that the steel companies who had produced steel for bridges then pushed its use for skyscrapers. The union became the first branch of the International Association of Bridge and Structural Ironworkers five years later. The employers on their side, formed the National Erectors Association. Between the two sides the closed shop was the main cause of dispute

FURTHER READING:

S. Giedon, *Space, Time & Architecture*, Oxford 1959.
Albert Bush Brown, *Louis Sullivan*, Mayflower 1960.
Louis H. Sullivan, *The Autobiography of an Idea*, Dover Publications 1956.

6   Building with Concrete

Buildings of concrete became more common in Britain during the first years of the twentieth century. The Royal Liver Building at Liverpool, claimed to be 'the most imposing example of ferro-concrete work hitherto undertaken in any part of the world', was near completion in 1909. The carpenters, bricklayers and masons began to feel the adverse effects of concrete on their work. Since the early nineteenth century concrete had been used only for foundations of public buildings and bridges and for harbours, jetties, and dry-docks as at Woolwich and Chatham. Its greatly increased use for whole structures was made possible by the invention of reinforced concrete and later, in the twentieth century, of prestressed concrete.

Many men in different countries, Britain, France, Germany, America contributed to these inventions. The Chief Figures, however, in the introduction of reinforced concrete, *béton armé* as the French call it, were all Frenchmen; Joseph Monier, François Hennebique, and Eugene Freyssinet.

JOSEPH MONIER (1823–1906)

A number of men were thinking about combining iron and concrete in the eighteen-forties. There were two main reasons for this. Before there could be a waterproof concrete which would prevent corrosion of the iron reinforcement there had to be improvements in cement. These came when Portland cement was invented by Joseph Aspdin. By that time there had developed a great demand for a reliable hydraulic cement. The industrial revolution in Britain was accompanied by vast construction of canals, roads, bridges, tunnels, docks and harbours. On 6th November 1824, the *Leeds Mercury* reported: 'We hear that Joseph Aspdin, bricklayer of this town, has obtained a patent for a superior cement resembling Portland Stone.' Aspdin described his method as follows:

'. . . I take a specific quantity of limestones such as that generally used for making or repairing roads, and I take it from the roads after it is reduced to a puddle or powder; but if I cannot procure a sufficient quantity of the above from the roads, I obtain the limestone itself, and I cause the puddle or powder, or the limestone, as the case may be to be calcined. I then take a specific quantity of argillaceous earth or clay, and mix them with water to a state approaching impalpability, either by manual labour or machinery. After this proceeding, I put the above mixture into a slip pan for evaporation, either by the heat of the sun or by submitting it to the action of fire or steam conveyed in flues or pipes under or near the pan till the water is entirely evaporated. Then I break the said mixture into suitable lumps, and calcine them in a furnace similar to a lime kiln till the carbonic acid is entirely expelled. The mixture so calcined is to be ground, beat or rolled to a fine powder.'

He himself had been influenced by John Smeaton, the builder of Eddystone lighthouse, who also had lived in Leeds. While Smeaton was experimenting with materials for the lighthouse he had discovered a combination of clay and lime which when pulverised and calcined had hardened under water. By the early 1840's Aspdin's son William had started a factory at Rotherhithe, on the Thames, to make his father's cement. *The Builder* reported in 1844 that this Portland cement had recently come into use in London for buildings, exterior stuccoing, chimney pots, copings, landings and pavements. At the building of the new Houses of Parliament the contractors, Grissell & Peto, carried out tests which showed it was better in strength, adhesion and appearance than the older type called Roman cement. The use of Portland cement spread, especially after the publicity given by tests made at the Great Exhibition of 1851.

When a waterproof concrete was possible, men's minds began to turn to reinforcing it. The uses of concrete itself were limited because it was resistant only to compressive forces. If the tensile strength of iron could be combined with concrete then it could be used in many more ways.

Others besides Monier led the way. Another Frenchman, Joseph-Louis Lambot, a gentleman farmer without any technical knowledge, made in 1848 some rowing boats for which he plastered concrete over a rectangular mesh of iron rods. They were for use on his estate in the South of France. Two of these boats can still be seen in museums in France. Joseph Monier made his beginning in the following year.

He was born in St. Quentin-la-Poterie, a village in the Gard district of Languedoc in the far south. He moved to Paris and became a gardener at the Palace of Versailles. While making wooden tubs for shrubs he tried cement but found it was too brittle. He then experimented with combining it with iron. Eventually he succeeded in making tubs and large pots with a mesh of iron rods enveloped in cement mortar. He developed his technique and extended its uses slowly so that it was not until eighteen years later that he obtained his first patent for reinforced concrete in July 1867. This covered portable containers and bowls or basins.

From that time Monier's progress was quicker. He had no scientific training but developed the uses of reinforced concrete by trial and error. During the next five years other patents followed. These were for tanks, pipes, footbridges and bridges. During these years, in 1872, he actually built a reservoir of reinforced concrete, to hold 130 cubic metres, at Bougival. After another six years he took his first step into building and construction when he took out a patent for a reinforced concrete beam. He became a building contractor specialising in water reservoirs and sewers. Two years later, in 1880, there was a serious earthquake in the French Riviera. By that time Monier's inventions were so much accepted that his methods were used in the reconstruction work. Thus like Paxton before him, the gardener Monier had become an expert in the building industry. But by that time other pioneers had been busy.

In England the invention of reinforced concrete for building, as distinct from making containers, came before Monier's. A Newcastle upon Tyne plasterer, William Boutland Wilkinson, took out a patent in October 1854 for beams or floors in which flat iron bars or wire rope were embedded in the concrete. Wilkinson had started his business thirteen years earlier and had then become acquainted with William Aspdin who had set up a cement factory in Gateshead nearby. Eleven years after his patent, in 1865, he built a house which was completely of reinforced concrete, including the precast stairs, the chimney and the walls which were 12 inches thick on the first storey and 9 inches thick on the second. He must have understood the part played by tension in the reinforced beams. For this he used second-hand colliery wire rope embedded in concrete. His ceiling and roof construction has been described by Dr. Hamilton as follows:

'The ceiling was made up from permanent shuttering of gypsum plaster forming coffers. The ribs were at about 26 in. centres and were 4 in. wide, with about an inch of the precast plaster on each side and about 2 in. of coarser plaster filling around and over the wire-rope reinforcement and up to the top of the shuttering. Over this was laid portland-cement concrete $1\frac{1}{2}$ in. thick with what we would now term a granolithic finish, each square bay being reinforced in both directions with two iron bars $\frac{1}{8}$ in. by $\frac{3}{16}$ in. in section.'

This house [which had been the home of a foreman of a builders' yard] was still in excellent condition when it was demolished in 1954 to make room for new laboratories at the Rutherford College of Technology. But Wilkinson's lead was not followed by others in Britain. Even thirty years after his house was built the experts in London could not see much value in a combination of concrete and iron.

Another pioneer in France had also been busy. Early on, in 1852, François Coignet, a civil engineer and manufacturer, built a house in Paris for which he encased an iron skeleton framework in cement concrete. In fact he took out a patent in England three years later, that is a year after Wilkinson's, which followed the same method. It referred to 'burying beams, iron planks, or a square mesh of rods in concrete or falsework'. He followed this up in 1861 by writing a report about the uses of concrete which had an armature of iron bars. He foresaw many of the modern uses but he could not describe them in detail or give designs because he had no theory of stresses. However, his son Edmond, who had the theoretical knowledge required, continued his work, and the Coignet system became a competitor of Monier's,

It was in the United States that a theory of reinforcement was put forward first. Thaddeus Hyatt (1816–1901), a lawyer of New Jersey, published in 1877 a book called *An Account of Some Experiments with Portland Cement Concrete Combined with Iron as a Building Material with Reference to Economy in Construction and for Security against Fire in the Making of Roofs, Floors and Walking Surfaces*. The following year he took out a patent for construction in reinforced concrete. He had made beams and tested them scientifically in New York and in England. He showed that bars were more economical than rolled beams as reinforcement; he established the ratio between the expansions of iron and concrete, and recommended 'the use of a deformed bar with bases or raised portions formed upon flat surfaces of the metal'. But when *The Builder* reviewed his book, it could only see the value of his discoveries for fire proofing, not for construction.

Hyatt was one of those men whose ideas fell upon stony ground. He was a liberal, public spirited man, a strong supporter of the abolition of slavery. He wanted the country to benefit from his discoveries but he never found an engineer or builder who would put them into practice. The situation in America was against him. Portland cement had to be imported into the United States at that time and for many years it was regarded as an expensive luxury. The manufacture of American Portland cement, in Pennsylvania, only began on a small scale at the time when Hyatt wrote his book. On the other hand, steel was more plentiful, and as shown in Chapter 5, was soon used in construction.

In America the first fully concrete-framed building did not appear until 1903, a quarter of a century after Hyatt's patent. This building, in Greenburg, Pennsylvania, was the work of an Englishman, E. L. Ransome (1844–1917). It took him many years to reach that stage. The son of the manager of Ransome and Sims's Ironworks in Ipswich, he had served his apprenticeship there. At the age of twenty-six he had emigrated to America to be manager of the Patent Concrete Stone Company in San Francisco. He became interested in the methods of building factories. First he was able to do away with cast-iron beams by using iron tie bolts embedded in concrete from wall to wall. Then by gradual stages of improvement he reached the point of patenting his own complete 'System of Unit Construction'.

Let us return to Joseph Monier in France. He had patents in Germany but he sold them and there they were developed theoretically and widely used. In France his system was replaced by others. Coignet's son developed his system, and a new force in reinforced concrete came to the front in the energetic person of François Hennebique. Monier himself died in poverty at the age of eighty-three, in March 1906. The men who had prospered in the industry had made a collection of money for him. A few months later a contemporary journal reported his death:

'Almost unknown, almost forgotten, in unfortunate circumstances, while many who used his ideas are among the best known and wealthiest practitioners in reinforced concrete.'

FRANCOIS HENNEBIQUE (1842–1921)

One of those who certainly was to prosper with reinforced concrete was Hennebique. He spread its use throughout the world. Hennebique was an Artesian, a native of the province of Artois in northern France, which gave its name to the artesian well. He was born at the village of Neuville-St. Vaast, five miles north of the city of Arras. The village which was just south of Vimy Ridge, the scene of bitter fighting in the 1914–1918 war, was destroyed at that time and completely rebuilt.

Hennebique's father was a farmer like his forebears before him. He expected his son to follow him and young Hennebique worked on the farm after leaving school. However, he had somehow developed an interest in scientific and technical subjects and he bought textbooks in mathematics, chemistry and physics. He spent his evenings studying them, after a long day's work on the farm, when the family were asleep.

In the meantime he got to know a building contractor in Arras and became interested in construction. The stone cutting and dressing he saw in the contractor's masonry yard and workshops interested him particularly. He told his father that he did not want to be a farmer, which his father accepted regretfully but with a good grace. He was then taken on by the contractor to be trained as a mason.

By working hard Hennebique cut short his period of training while at the same time he continued studying technical subjects. Much of his work was in building churches and restoring the old ones scattered over the plains of Artois. One of his jobs while he was employed by the contractor was to supervise the construction of a church. By the age of twenty-five he was able to start his own business. This he built up by church restoration work. In order to understand the subject thoroughly he studied archaeology and visited many of the cathedrals of France so as to become familiar with the old Gothic style and methods of masonry. Before long he had such a reputation for restoration throughout northern France and Belgium that there was more work than he could deal with. Two outstanding examples of this work were the Churches of Notre Dame and of St. Martin at Courtrai (now Kortsijk) near Ypres, in Belgium.

After a few years, ambitious to expand his business, he undertook contracts for public works, railway works, viaducts and bridges. Responsible now for work in concrete and steel, rather than stone, he began to think about the possibilities of combining these two materials in a new way. He was led to invent his own system of reinforced concrete by a demand that the floors of steel girders and concrete arching which he was providing should be made fireproof. It was the desire for a type of construction which would resist fire better than the ordinary steel and concrete of that day which led to reinforced concrete in his case.

In 1880, the same year as the earthquakes on the Riviera, he took out a patent for floor slabs reinforced with round bars. Soon afterwards a client for whom he was designing a house asked for it to be made fireproof, and accordingly the complete floors were made of the patent slabs. This house was at Lombardzeyde, a village on the Belgian coast near Westende; it was destroyed by artillery fire in the 1914–1918 war.

Hennebique always stressed the advantage of fire-resistance when he was selling his system and he published a booklet, *No More Disastrous Fires.*

His next technical step was in 1892 when he patented his (and the first modern) type of concrete beam with round main bars fishtailed at the ends and vertical stirrups of flat hoop-iron, and about this time he began the use of 'T' beams for floors. His mix of cement, sand and stone was about 1 : 2 : 4. In that year, now fifty years old, he also took another important step. Feeling confident that his system of reinforced concrete was practicable and safe, he decided to concentrate on promoting its use. He gave up his business as public works contractor and started a new firm, Maison Hennebique (The House of Hennebique) solely for construction of buildings using his patents and methods.

The Maison Hennebique grew rapidly during the eighteen-nineties. From small works he went on to large ones, his reputation increasing all the time. In 1894 he built the first railway bridge, a small one with a span of 2·40 metres in Switzerland, and a spinning mill at Mulhouse; in the following year a large granary at Lille; in 1896 a cantilevered quay at Nantes and in contrast, the underpinning of the tower of Notre Dame de Brebières at Albert. This tower, weighing 8,870 tons, had quickly sunk five inches and Hennebique was called in to provide a reinforced concrete raft. In the following year he patented an important invention which he had worked on for two years—and which made a sensation—piles of reinforced concrete. An engineer on his staff wrote:

'When M. Hennebique came one morning to the drawing office and asked his chief engineer to design a ferro-concrete pile, sketching himself the external lines and internal reinforcement, and had left the room, all the engineers and draughtsmen commenced to laugh, saying the patron now wanted to make everything of ferro-concrete. However, when the required drawings were finished, M. Hennebique brought a sketch showing the details of a driving helmet made of steel plate, but otherwise on nearly the same lines as the cast steel helmets employed in the present day.

'The staff then began to realise that the ferro-concrete pile might not be so impracticable after all, and when the first pile had been successfully driven, their doubts were finally dispelled.'

In 1898 and 1899 he built two bridges. The first a footbridge with 15 metre span over a railway cutting at Estemay in the Marne district; the second a much bigger operation, was the bridge at Châtellerault across the river Vienne which flows into the Loire. This with its length of 144 metres, and one span of 40 metres and two of 40 metres, was a major task.

During that decade of the 'nineties Hennebique had organised his technical staff, and appointed agents who selected his licensees. As a result his firm had carried out some 3,000 structures in reinforced concrete of a total value of about £2 million. He developed a highly organised business which promoted his system all over France and Europe. He issued a monthly journal, *Reinforced Concrete*, and held annual conferences of his licensees and employees. At one of these, in 1899, a song enlivened the proceedings:

'Hurrah for reinforced concrete!
We are all sons of the same father
Friends let us make the system prosper
Let us united pray
That all the world over
Our big family grows from day to day
And that in the whole universe none other
But reinforced concrete is used from this day!'

By this time it was necessary to move the business to Paris, the financial centre of France. Therefore in 1900 Hennebique built there, in the Rue Danton, a large office and residence. This building of eight storeys was constructed entirely of reinforced concrete; floors, walls, stairways and ornate front. The special permission of the city authorities had to be obtained, because of the new methods of construction. The house in the Rue Danton became a training centre for hundreds of young engineers, who were sent out to agencies and offices which had begun to spread beyond the continent. 1900 was a good year for Hennebique. He exhibited at the Paris international exhibition and was awarded a grand prix, and built a large bridge across the Loire at Decize with two arches of 185 and 165 feet spans.

The Hennebique system had already been introduced into Britain. In 1897 he had taken out a British patent which said:

'. . . by arranging in a mass of beton of suitable form, longitudinal bars of iron of a given shape in order to constitute the tension chord, by distributing them in the mass in a judicious manner in order that the whole mass of iron and beton may have at every point of the piece formed the desired resistance to flexion and breaking strain, and by further connecting the longitudinal bars by brace pieces or stirrups of suitable form, I have succeeded in producing the practical joists, girders and the like which form the object of my present invention.'

In the same year a Frenchman, Louis Mouchel, opened an office in Victoria Street, London, staffed by French engineers, to carry out building in reinforced concrete under Hennebique's patents. The first building, a flour mill in Swansea, was followed in 1899 by a jetty or wharf at Woolston near Southampton, 136 feet long and 46 feet wide. He used a well-graded dry concrete mix.

Hennebique took out many other British patents. Many other patent systems followed which differed from his only in the form of the bars. But in spite of competitors, chief among whom was Edmond Coignet, his system held the field in Britain for some years. The Royal Liver building in Liverpool, under construction in 1909, was designed by his agent Mouchel. However, the use of reinforced concrete in Britain developed much more slowly than in the continent or in America. The building regulations were partly responsible for this. Mouchel said in 1904 at a meeting of the Royal Institute of British Architects that London enjoyed 'the unique privilege of being the only city in the civilised world where reinforced concrete constructions are actually prohibited'.

A few large British firms, Cubitt and Company, and Holland and Hannen, did try to promote the new material. But, although England was the home of Portland cement, the engineers and architects were too conservative to welcome it. Progress was made slowly after the Concrete Institute (later the Institution of Structural Engineers) was founded in 1908.

In the meantime in France, state regulations for reinforced concrete had already been issued. Hennebique was a member of the Government Commission which had been set up to report on the industry. But he expressed his opinions of the intention to draw up official methods of calculation and regulation in no uncertain fashion. Holding up a blank sheet of paper he said: 'If we make any more regulations than this we shall be criminals and we shall strangle all progress'.

The Maison Hennebique continued to expand rapidly all the same.

It constructed railway, road and river bridges, reservoirs, warehouses, harbour and docks, factory buildings over a vast area—Germany, Russia, Britain, Italy, Egypt, Turkey and the U.S.A. between 1900 and 1914. One notable structure was a caisson 78 feet long by 55 feet wide by 71 feet high which when sunk in the roads of Hyères was used as a torpedo testing battery in the Mediterranean. Another was the Risorgimento Bridge across the river Tiber in Rome with an arch to span 100 metres, a record for many years, and a rise of only 10 metres.

Hennebique became a very wealthy man. Continuing to live in the Rue Danton he also built for himself a reinforced concrete villa on the southern outskirts of Paris and had an estate in Normandy.

EUGÈNE FREYSSINET (1879–1962)

The modern method of prestressing concrete is due to Marie Eugène Freyssinet. He was born in the village of Objat on the Corrèze plateau in south central France. It was a region of rocky wastes, thick forests, poor soil and a hard climate. He came of peasant stock. In his own account of his life and work he wrote:

'For many centuries my ancestors lived hanging to the sides of deep gorges from which the streams of Corrèze flow. These men were versatile craftsmen who made for themselves a way of life based on great care for simplified forms and economy of means. They developed a style of construction, of art and of life which in many cases reached the level of a masterpiece.'

His grandmother, who was born in the mountains, worked an ancient flour and oil mill. His father, an orphan from his birth, was at first an agricultural labourer, but by dint of his own intelligence and energy made himself into an educated man with a liking for the arts. When Freyssinet was six his parents moved to Paris, taking him and his younger brother, who later became an artist, with them. He was sent to the parish school in the Rue des Ecluses-Saint-Martin. He liked the teachers there who were good at their job but his school mates were another matter. The rowdy Parisian children teased him unmercifully because of his provincial accent and awkwardness. After many fights they left him alone and from then on he lived in his own world of dreams which he created to shut out the Paris he detested but where he had to live. He had one illness after another so that his parents were obliged to send him back to the Corrèze country for long holidays, in spite of the handicap to his progress at school.

During these holidays he had his real education. This he received from the craftsmen of the district, carpenters, joiners, smiths, weavers; they were his friends rather than any children of his own age. The great event was when the mill stopped working after the harvest. The miller put out nets in the river and opened the sluice gates. Some fifty countrymen with the picks and shovels arrived to repair the watercourse. Then they all sat down to consume the fish caught, together with the hares and partridges killed in the fields. The mechanic, the carpenter and the mason came to do repairs to the mill and lived with the family. There was always a wheel to repair, a millstone to be seen to, an embankment to be strengthened, a barn to be enlarged, and young Freyssinet watched everything. He wrote:

'These men were my first and most effective teachers, who made the strongest impression on me. Thanks to them I was a complete craftsman at twenty years old, able to find the best solution to any of the numerous problems in the mill or the farm and to help put it into practice with my own hands. However, the greatest lesson those masters of my youth taught me was the example of their spirit which united in me the instincts for building which I had; and which gave me such a strong feeling for technical honesty that now I resent any lack of professional conscience or courage like a kick in the behind. What I learnt from them was the willingness to accept a given task, the beauty of the craftsman's work, and the value of every determined and persistent effort.'

Back in Paris he discovered the museum of arts and crafts, near his home where, as a boy of eleven or twelve, he attended evening courses on chemistry, physics and electricity. With a few pence he bought old apparatus in the market and tried to make all kinds of unlikely assemblies. Happy in his room he was miles away from the Parisian children.

At the age of twenty he was enrolled at the Polytechnic, having failed to enter the year before. He liked the atmosphere there, the lecturers, and the serious hard-working students most of whom were from the provinces. The one unpleasant aspect was the system of discipline which he found unbearable. However, he passed out in such a high position that he could choose his future career in the Department of Bridges and Highways. At the Polytechnic he had never quite accepted the belief, current there, in the supreme power of mathematics; his craftsman's instincts were too strong.

During his year of military service, as second lieutenant in the corps of engineers, he made his first invention, a system of ferry-boat equipment which later became regulation. Then he joined the school of the Department of Bridges and Highways. It was there that the idea of prestressing reinforced concrete first came to him, under the influence of one of the teachers. By that time the theory of reinforced concrete had advanced so far that this teacher could give a course on it. This brought home to Freyssinet the qualities and the shortcomings of the material and also the idea of stressing. The possibility of bringing these two together and of producing 'permanently active forces in the reinforcement', began, as he said, 'to haunt my mind between 1903 and 1904'.

In 1905, then twenty-six years old, he entered the government service and was appointed bridge and highway engineer at Moulins, which was on the river Allier, a tributary of the upper Loire, and the chief town of the Allier Department in central France. He was responsible for local communications and technical adviser to the mayors of the district. He knew the needs of the countrymen, above all for bridges to take the place of fords. When the first one came to see him he promised that if the mayor would provide the earthworks and transport he would build him a bridge wider than the regulations provided for and at 20% of the official estimated cost. This he did, and a number of others in the next few years, economising mainly by reducing the size of the expensive metal platforms which had to be brought up the valleys.

'The design had to use the local conditions to the best advantage. I arrived at it after pondering the problem for a long time, making innumerable sketches with complete freedom from any precedents. After a hundred attempts an idea sprang from my subconscious which I realised was the only possible one. I then designed the work in the smallest detail, believing that it is the details which make work good or bad. I carried it out either by direct labour teams which I had formed or by small local builders whom I had completely in hand. I was also the carpenter, coffer maker, iron worker and cementer. Wielding my axe, hook and trowel, I taught the lads of the village, or rather we learnt together the best and quickest ways of making coffers and armatures and of pouring cement.

'Naturally the mayors of my district approved highly of me and if anyone had told them that my activity was against the regulations it would have been at great risk to himself. All the same it would have been true; no authority had approved my projects on which I had ignored the official circulars.'

A young man of twenty-seven, he was happy making these small low-cost bridges for the local towns and villages and, as he was not ambitious, he might have continued in that way. he met one of the big men in the construction industry, François Mercier, who put bigger things in his way. It so happened that in 1907 there were three suspension bridges each about 250 metres long over the river Allier which were due for replacement. Only one of them, at Boutiron, was in his district and it was not likely to be reconstructed for many years. However, he designed a new bridge and, as money was tight, at the lowest cost possible. At that point the river ran over drifting sands which made it necessary to drive piles into the rocks below. As piles were expensive Freyssinet designed only two which meant three arches of $72\frac{1}{2}$ metres span. For the same reason he included the minimum amount of steel and of concrete, and pinned up the design in his office. His friend Mercier was so taken with it that he guaranteed to the government the cost of all three bridges provided they were built to Freyssinet's design.

These three bridges took Freyssinet into a bigger world of construction. They occupied his life from 1907 until 1912. They were big jobs. He followed them up in 1914 with another bridge over the river Lot at Villeneuve which had a single span of 97 metres. While busy with the problems of construction and while grappling with them he noticed certain features of the reinforced concrete he was using which led him later on to his discovery of methods of prestressing. In fact he had already constructed a large prestressed tie member and tested it. But it was a mishap with one of his bridges in 1912 caused by shrinkage in the reinforced concrete which made him realise the existence of changes in the material far greater than those allowed for in the official regulations. One day he realised that his bridge at Le Vendre on the river Allier was threatened by disaster:

'It seemed to me that the lines of the parapets which had been perfectly straight were very slightly and slowly becoming convex towards the sky . . . when I verified that the movement was increasing and more quickly my uneasiness became acute anxiety. . . . I was then certain that my arches were becoming deformed vertically. . . .

'During the night I jumped on my bicycle and hurried to Le Vendre to wake Biguet (the foreman) and three reliable men. We five applied centring jacks so as to counter the deformation which had developed. As soon as there was light enough to use levels and poles we began to lift the three arches at the same time. It was market day; frequently the operation had to be stopped to let vehicles pass. However it all ended well: its lines restored, cured of the weakness which had almost destroyed it, the Le Vendre bridge behaved perfectly until it was destroyed during the war in 1940.'

This mishap showed Freyssinet the existence of creep in reinforced concrete and he immediately started research to discover its laws and ways of counteracting it. This led him after many years to prestressed concrete. He carried out the research at his home and on sites. In his garden he had a mechanics laboratory equipped with apparatus for measurement which he made himself. There a young and enthusiastic team joined him in the study of creep, using an experimental arch. But in August 1914 they were all called up for active service. The piece of ground where he had his arch, apparatus and notes was handed over to the military authorities and when he returned in 1919 he found nothing but a slag heap. All the work he had done was lost.

For the first three months of that war Freyssinet was stationed on the Alpine frontier as Italy did not ally with France until 1915. Then his specialised skill was called for and he became engineer officer of military works in the north. Steel was scarce and the demand was for reinforced concrete. Accordingly he constructed many cargo sheds, factories, hangars, boats and howitzer carriages of concrete.

When peace came he left the Government service and with some associates formed a large firm specialising in construction in reinforced concrete which for many years was busy in the devastated areas of France. As a director he was responsible for technical development. Eventually he had the time to start again his earlier research into the nature of creep and shrinkage in order to see if permanent prestressing was practicable. The stimulus came from a serious case of cracking in the main beams of the Caen steel works in 1926 which he was asked to investigate. He started experiments which lasted three years. At the same time a great bridge he designed gave the opportunity to study the movement of the reinforced concrete used.

This road and rail bridge was the famous one at Plougastel over the river Elorn, where it emptied into Brest harbour, which was completed in 1930. It had three arches, each of 612 feet span, a record

size until 1934. While it was under construction Freyssinet wrote a
paper in 1929 about it. Here are some extracts:

'The river at this point is 2130 ft. wide at high tide. The tide attains
a height of 26 ft. and the current is occasionally very strong. The
minimum opening in a bridge was required to be 564 ft., with a clearance of 118 ft. above a channel in which not even temporary support
were possible. The present design for the bridge was chosen in a
competition because of its economy and durability. Altogether much
cheaper than any of the competing designs, it permits the carriage of
both a highway and a standard gauge railway, whereas the other designs
provided only for the highway.
   'A great arch being obligatory for the spanning of the channel, I
judged it economical to use for the remainder of the crossing two
identical arches in order to re-use the centering. The work then
comprises these three arches of reinforced concrete of 612 ft. span
centre to centre of the piers. They support a deck of two storeys, the
lower of which carries a single track standard gauge railway, and the
upper a roadway 26·2 ft. wide.'

So far as the arches were concerned the amount of steel employed
was only about 37 lb. per cubic yard of concrete.

'The concrete employed for the arches contained on an average
1·9 bbl. of ordinary portland cement per cubic yard of finished concrete
with an aggregate formed of 4 parts crushed quartzite, 1 of sand residue,
and 1 of bar sand, and gives strengths which will easily attain 8500 lb.
per sq. in. at the time the bridge is placed in service.
   'The arches were poured by means of electric cranes on a cable, by
a new system constructed by ourselves, of 2260 ft. span on a centering
consisting of a wooden arch of the same span as the concrete arches
and taking its support on their foundations.
   'This centering was constructed on the shore and transported to its
position of use for the first arch; taken away forty-eight hours after
completing the pouring of that arch and put in place for the second
arch, from where it will be transported to the third arch. Four months
elapsed between the first and second moving of the centering.
   'The second placing of the centering was accomplished in spite of
a very strong wind, without the slightest difficulty, in less than 3 hours.
   'No bolts or ties are used in the centering. The work is reduced to
strokes of the saw for which no precision is required, and to nailing.

The nails used are of the ordinary kind except for pegs without a head 0·4 in. in diameter and 14 in. in length. Eighty-eight tons of nails were used. The framework was constructed for the most part by unskilled labour and on land, profiting by the favorable shape of the bank.'

Already in 1928 Freyssinet was using for it a form of prestressing. One arch was destroyed in the second world war and restored five years later.

Before the Plougastel bridge was finished Freyssinet had finished his research. While it was being built he had written:

'Concrete reacts and adapts itself to the deformations as a living being. It shrinks locally in order to avoid excessive strains and to carry them back to zones under less strain.

'In an arch formed by successive barrels, the first barrels taking more important contractions than the last barrels, the inequalities of pressure between the different barrels tend in the long run to lessen considerably.

'These results are important for the construction of large concrete arches and it is desirable that the experiments be made again and carried on in other climates with other cements and other proportions.'

He was now able to make a quantitative assessment of creep and shrinkage and hence to say that a permanent prestress in concrete was practicable if high strength steel at a high stress was used. In other words, it was necessary to use steel of such high strength that sufficient tension remained after the shrinkage and creep had taken place. He wrote:

'As soon as I knew the laws of slow deformations well enough to be sure that they could be reconciled with permanent prestressing I decided to risk my living, my reputation and energy to make the idea of prestressing an industrial reality. . . . At 50 years of age I relinquished a life marked out before me for a new one full of hazards and dangers.'

He felt he had a mission to fulfil whatever the risks. He had taken out patents covering his methods of prestressing but his co-director refused to agree to develop them. Therefore he resigned very reluctantly, giving up both financial security and the certain pleasure he enjoyed in his work as a structural engineer and builder in order to work out the uses for his patents.

He took the first step. The first licensee of his patents was a company connected with the electricity supply in the Paris area in which he had many friends. He turned an old depot into a factory equipped with machines to produce with accuracy large poles 16 metres long of prestressed reinforced concrete. Their weight was only 40% of that of standard poles, the steel in them was only one-third of usual practice, and they were much stronger. Freyssinet wrote:

'The years of creation of these techniques: 1929 to 1933: were for me the most intense activity which I have ever known. How many anxious nights were passed while I and my colleagues, whose faith and emotions were as great as mine, watched the result of an experiment. I felt I was creating new techniques of the greatest importance.'

But though he was successful technically the enterprise failed. There was not sufficient demand for these transmission poles. He had lost all his money. In fact, however, the disaster was a blessing in disguise. If the enterprise had succeeded he might have continued with the same kind of articles and bigger ones so that it would have taken many years to prove his methods and reach to the scale of large public works. As it was, by chance, he was able to seize an opportunity in 1933 which took him straight away into large-scale works.

The new Maritime Station at Le Havre was slipping into the harbour. Because of settlement in the bed of mud under the foundations the building seemed doomed to destruction on the eve of its opening. Freyssinet put forward a scheme to save it which involved many applications of prestressing, the rapid use of techniques which had never been applied to this purpose before, and the creation of new techniques. This was for him a unique chance to publish the techniques on which he had spent five years' effort and his money as well as to restore his confidence in himself and the value of his work. Delay was dangerous; his scheme was accepted. He wrote:

'I claimed to be able to stop these settlements and to save the building by using this new process of prestressed concrete. Since there was no choice between ignominious collapse and my own scheme with all its daring features, the smallest of which would, in normal times, have sufficed to rule it out of court, my scheme was adopted, and it had the good fortune to succeed. The risk had been great. But I had broken that vicious circle which encloses all innovators, especially where public works are concerned . . . every authority requires reference to previous work. Thus was prestressed concrete launched.'

In fact as soon as the first piles were sunk the most dangerous settlements stopped and success was in sight. The operations attracted many prominent men, among whom was the head of a large construction firm who from then on backed Freyssinet in the exploitation of his patents. With the help of a new and expert team Freyssinet's invention gradually became accepted. It was used in building dams in the French colony of Algeria and docks at Brest. Special equipment for stretching and anchoring the steel wire and moulds for casting had to be developed and Freyssinet himself invented tools for stressing wire. His methods came to be used all over the world.

The first world war had destroyed Freyssinet's research; the second one stimulated greatly the demand for prestressing. In Britain it had not been used at all, but during the war engineers had to find a substitute for timber for railway sleepers and were obliged to use this foreign technique. On the Continent shortages of steel speeded up its use.

After the war he continued active in design. Prestressed concrete came into its own in a series of great works, bridges, dams, viaducts in many parts of the world. Britain made up for its pre-war neglect when he received a doctorate of science from the University of Leeds, an honour which the Belgian universities of Brussels and Ghent also conferred. France made him a commander of the Legion of Honour. He married but had no children.

Towards the end of his life Freyssinet wrote about himself and his work:

'Young people, do not believe those who tell you that I have had an exceptional career because I had exceptional gifts of intelligence. That is an excuse which lazy and weak people find for themselves. For what I have done I never needed an exceptional intelligence.

'The idea of prestressing is a simple one. To compress reinforced concrete so as to make it capable of resisting permanently ulterior movements is an obvious idea. My discovery of the rapid hardening of concrete? One did not have to be so clever to realise that in dividing by a hundred the value of the space between the particles at the beginning of the job one reduced in the same proportion the volume of hydrates necessary for hardening and also the time required. . . . The idea, however astonishing it might seem to the mathematician who only sees nature through a mist of x's and y's, was simple for a craftsman who felt the material with his fingers and had made joints and smoothed plaster with his own hands.

'Certainly to realise my ideas I had to have patience, perseverance and complete technical honesty. Perhaps if I had had those intellectual abilities which have been attributed to me I might have attained my object with less effort. Instead, I had untiring tenacity, and at least three times in my life was bold enough to scorn great risks.'

THE BUILDING WORKERS (1885–1914)

During the last years of the nineteenth century and the first years of the twentieth the building workers of Britain had their ups and downs. These were brought about in three ways: the changes in the capital invested in the industry, the actions of the unions and of the employers, and the effect of new methods and materials, such as reinforced concrete.

As always in building, boom and slump followed each other. The bad years of the eighteen-eighties led to great expansion of the industry throughout the eighteen-nineties, which in turn led to a ten year depression from 1900, ending with a slight improvement in the few years before the outbreak of the 1914–1918 war.

In the building boom of the 'nineties the London suburbs, for instance, spread rapidly. Ilford, Walthamstow, Enfield, Edmonton, Tottenham, Willesden, Ealing, Acton, Wimbledon, Croydon all grew by more than 30% during the decade. Labour was in great demand. The number of men and boys in the industry increased by a quarter of a million, or by about one-third. Unemployment was very low at about 2%. The trade unions raised their membership by leaps and bounds; the carpenters and joiners from 25,000 to 65,000, bricklayers from 7,000 to 39,000, plumbers from 4,000 to 11,000, stonemasons from 10,000 to 19,000. This meant, in fact, that a much greater proportion of craftsmen were members of trade unions than before. For building as a whole this proportion rose from 10% to 19%, and for carpenters and joiners from 13% to 25%. This strengthened the hand of the unions greatly, though employers sometimes still insisted on non-union men being represented in negotiations, as the Central Association of Master Builders of London did in 1890.

In these years the unions' strong position was not weakened by changes in materials. Steel and reinforced concrete did not come into force until the twentieth century and their use did not displace craftsmen as yet. All the time, however, brickwork was replacing masonry.

It was inevitable in such circumstances that the workers should advance; this they did in many ways, but not as they might have done. They were able to push up their wages by about 15% during the ten good years, reduce hours, and strengthen craft practices. On the other hand, the way they did it weakened their unions. Branches and local federations of crafts often made and fought their own claims without control from their head offices. In May 1891, a united trades committee of carpenters and joiners in London struck three firms for an increase in wages and an eight hour day. After six months the 4,000 men involved gained a reduction in hours from 52½ to 51¼ and better rates for overtime. This was the first of many strikes in the next few years, many of them unofficial. That particular strike led to a federation of all London building trades, but these local bodies were on bad terms with the headquarters of the unions who were more cautious and so the national unions were weakened.

Another feature of the building unions during the 'nineties was that they took hardly any notice of the turmoil and the great forward movement in the trade unions generally. The appearance of the big new unions of unskilled workers, the dockers' strike of 1889, the rise of socialism, the beginnings of the Labour Party, all passed them by. In fact their representatives on the T.U.C. resisted the new wave of militants as long as they could. The building workers who had been on the left wing of the labour movement were now on its right.

The employers were in retreat; short of labour, they often gave way to local demands without a struggle. However, they strengthened their organisation. The National Association of Master Builders (today the National Federation of Building Trade Employers) grew rapidly as more and more local associations such as the London Master Builders joined it. In it were some 1,300 firms, all contractors, as speculative builders were not respectable enough to be admitted.

Conflict between the two sides arose mainly from two issues, apart from wages and hours. Piecework had been condemned for many years by the unions because it meant bad workmanship and lower pay. The Amalgamated Society of Carpenters and Joiners had encouraged its branches to resist piece-work and in 1891 made a rule banning it completely. Stonemasons, bricklayers, plumbers and plasterers followed suit. As piece-work usually meant sub-contracting the hiring and supervision of labour, the unions also acted against this. For instance they persuaded certain public authorities, such as the London School Board which was responsible for elementary schools and the London

County Council to include in all their contracts a clause forbidding any sub-contracting of labour.

But a more important issue was apprenticeship. Most of the crafts made strong efforts to maintain it, particularly the plumbers. The exception was the painters; almost anybody could become a painter, e.g. 'labourers, sailors, fishermen and broken-down gentlemen's servants', as the *Painters Quarterly* said in 1896. The employers, faced with a shortage of labour, wanted to remove limitations on the number of apprentices but were unable to do so. Sometimes the limitation had been accepted as one of the agreed working rules, and in Nottingham employers had been fined up to £50 for breaking it. Therefore the employers began to act against limitation which they said must 'trench upon the just and inalienable rights of the employer and would be a natural calamity'. By 1898 they had started companies in several towns 'for the removal of this restriction which is severely handicapping employers generally'.

This question was among those which led to the conflict in 1899 when a national lock-out of plasterers, affecting about 3,000, lasted for several months. 'The plasterers', complained a Leicester employer, 'were probably the most impudent and independent class of men in the building trade'. Another in London declared that 'the masters were in the hopeless position of being obliged to accept almost any terms that the Plasterers made with them'. The plasterers' union, the National Association of Operative Plasterers, had grown rapidly from 1,470 members in 1887 to just on 11,000 in 1899. It had a new secretary, Martin Deller, who wrote of himself: 'I have the reputation of being somewhat of a fighter', but whom others called forceful, domineering, overbearing, intolerant.

The employers' knowledge that the boom was coming to an end and the fact that plasterers were specially scarce led them to attack the N.A.O.P. First the London contractors tried to find a way round the ban on controlling labour by sub-contracting it, by appointing 'managing foremen' with extra discipline. The union countered by making the foremen become members and when some refused there was an unofficial strike. The employers then decided 'to deal with the whole question of plasterers throughout the country' and demanded an end to strikes against non-union foremen, to the boycotting of non-union workers, to limitation on apprentices. When plasterers did not accept these a national lock-out began on 6th March, 1899.

Both sides stood firm: the employers declared 'the so-called master has for years provided the brains and capital, only to be robbed and plundered by the ever-recurring restrictions and octopus clinging tentacles of the N.A.O.P.' Martin Deller wrote to them 'that nothing but a fight would appease your desire for annihilation'. A certain noble lord wrote to *The Times* warning the plasterers that they would soon be ousted by 'truly artistic Japanese woodworkers', paid 7d. a day on which 'they manage to live and thrive, though their food consists only of rice daily and two or three times a week the heads and tails of fish; they cannot afford to buy the bodies also'. All the other building unions came forward to help the plasterers in a unity that had not been seen for many years. Eventually on 30th May, the lock-out ended when the plasterers agreed to end their action against foremen and non-unionists, and the apprenticeship dispute was shelved.

It was typical of the industry, however, that at the same time as this conflict there was also more conciliation and arbitration in some areas and some trades. Only three years before the plasterers' lock-out the Master Builders' Association in London and the unions set up a local conciliation board. Then in 1897 the plumbers achieved a national conciliation board which aimed 'to promote and secure, if possible, an honourable and equitable adjustment of any matter or question pending between employer and employed with a view to avoiding or preventing strikes, lock-outs or other measures which prove disastrous to out mutual interests'. For the unions and their members conciliation was the best way out of strife and when the plasterers' dispute was ended in 1899 they suggested a national conciliation board. But the employers demanded financial guarantees that union branches would keep any agreements made. As the unions could not agree to this the scheme was then dropped but it was soon revived with the result that five years later the first national machinery for more than one craft was agreed by the bricklayers, masons and carpenters and the National Federation of Building Trades Employers. Even the labourers were brought in after a few years, though not on an equal footing. However during the next ten years the new conciliation boards were not much used, mainly because the position of the unions became weaker.

Those years were the years of slump in building from 1900 to 1910. Investors found it more profitable to send their money overseas and develop the British Empire rather than use it to increase the stock of houses in Britain. Speculative builders were discouraged by having to pay a higher rate of interest on loans than in the eighteen-nineties. Overcrowding remained the same; in London it got worse.

Unemployment was heavy. The percentage of unemployed among the carpenters and plumbers rose to 8% in 1905 and 12% in 1909. All the unions lost members. Those that paid unemployment benefit such as the plumbers and the carpenters kept theirs fairly well, but the bricklayers' membership fell 40% and the stonemasons worst of all by even more. The proportion of all building craftsmen who were members of unions fell disastrously from 19% to a mere 13%. Matters were made worse by the impact of technical changes. The most important was the increasing use in Britain of reinforced concrete or ferro-concrete, arising from the activities of François Hennebique. Although it was not yet much used for houses, reinforced concrete replaced masonry and to some extent brickwork on most large buildings such as hospitals and hotels. The mason had to exist on small work and stone-cutting. The new material was limited mainly to floors and roofing because its use for walls required a special licence. Hence the carpenters lost a lot of work. The gloomy contemporary view was that 'the carpentry trade is doomed'. The plumbers also suffered from the use of asphalt concrete on roofs instead of lead. The concrete work was lost altogether by the building unions because the concrete labourers were a race on their own, not recruited from the building labourers or from craftsmen who had been displaced.

A second technical change, that of machines making woodwork, made many joiners unemployed. The secretary of the General Union of Carpenters and Joiners wrote in 1905:

'A large quantity of Joiners' work is being manufactured under worse conditions than has existed during the past 40 years. You will have gathered from our Organisers' Reports that these machine joinery works are now established in almost every district they have visited, and about the only places in which work appeared to be brisk, and with the exception of two or three leading hands and machinists, the remainder are underpriced men and youths with no recognised working rules or standard rate of wages. With trade drifting in this direction, and the substitution of iron and concrete and other materials in place of wood in the construction of buildings accounts to a large extent for the number of unemployed.'

These were not the only changes in materials and methods. Less use of 'compo', cement and sand, on house external walls affected the plasterers, as did also the manufacture of fibrous plaster slabs. The plumbers were hit by the use of earthenware for sanitary fittings, made by Doulton's and other firms, of iron piping in place of lead, and by the growing use of electric light. However, the Electrical Trades

Union, recently formed, did begin to co-operate with the building workers. Unfortunately the building unions were unable to meet the challenge of all these changes, because of their traditions, conservative outlook and inability to alter their constitutions and rules to draw in the new kinds of worker. This was the price they paid for isolating themselves from the new militancy of the trade union movement.

They could not, however, be independent of the sensational event which arose from the growing strength of the trade unions during the 1890s. This event was the Taff Vale decision of 1901. The railwaymen on the Taff Vale Railway had come out on strike. Their union, the Amalgamated Society of Railway Servants, recognised the strike and tried to run it in an orderly way. But the railway company sued the union for damages and was awarded £23,000 by the court which, on appeal, was upheld by the House of Lords. This legal decision was a great shock to the trade unions because they had believed ever since the Trade Union Acts of 1871 and 1876 that they were not liable for damages arising from a strike. Now their ability to carry on a strike was destroyed since they could not afford to pay heavy damages after each one: their only solution was to get Parliament to change the law. This came before long in the Trades Disputes Act of 1906.

The building unions reacted slowly to this situation by supporting the move to increase the number of labour members in Parliament. The T.U.C. had set up a Labour Representation committee, which later became the Labour Party. Gradually, and especially after the Taff Vale case, most unions affiliated to the L.R.C. but most of the builders did so only after a year or two.

The employers also reacted to the situation, but mildly, considering their power during these years. The National Federation of Building Trade Employers wrote to its members:

'. . . that it would be much better policy to aim at freeing the trade from such restrictions as paying incompetent men full wages, limitation of apprentices, and interference with business management, than attempting to lower wages. This Council strongly urges all Local Associations to aim at attaining these objects rather than attempting to lower wages.'

In fact although there were some reductions in wages, they never affected more than a small minority of workers. One reason for this was probably that prices and the cost of living were rising. Nor did the employers launch an attack on craft rules. There were much fewer strikes and lock-outs. Both sides seemed to want to keep quiet. One of the leading employers said, 'Between 1901 and 1911 we have had really a lovely time; it has been very quiet.'

The unions had other reasons to keep it 'very quiet'. Quarrelling between them increased enormously as the total amount of work available shrank. Also the technical changes made it harder to draw the line between kinds of work. There was conflict all the time between plasterers, bricklayers, masons and slaters about their share of the jobs going. Agreements were sometimes arrived at but with great difficulty. Here is a typical one:

'Agreement made between the Carpenters' and Joiners' Societies and the United Operative Plumbers' Association upon the question of Fixing Iron Gutters and Stack Pipes in the London District:
  1. That on all new work, all iron rain water gutters that are fixed to wood shall be done by carpenters.
  2. That on all new work all rain water pipes shall be fixed by plumbers.
  3. That the repairing of all rain water gutters and pipes shall be done by plumbers.
Repairing work is defined as applying to all roofs where the whole of the old gutters are not taken down for new ones to be substituted.'

The plumbers were the worst offenders. They had long and bitter quarrels with not only carpenters but the numerous trades who were taking some of their work and others—zinc workers, whitesmiths, glaziers, brassfinishers, slaters and plasterers. Their particular enemies were the gas fitters and hot water fitters with the spread of hot water systems.

It is not surprising that they clung to the work they had when one reads the following account published in 1894:

'At the age of twelve he was glad to leave school, and start work as an errand boy. His mother now died, and he went to live with an aunt, and engaged himself at a low wage to a plumber.

'Having a natural liking for this work, and thinking his want of education would not seriously impede him in it, he deliberately chose it as his trade, soon picked it up, was entrusted with skilled work, and stayed with the same employer nearly seven years.

'He was now thrown out of work, through a quarrel with his foreman, and could get nothing to do for a fortnight. Having, however, saved £10 or £11, he married during the fortnight within a few weeks of his aunt's death.

'Since his marriage, the man's employment has been marked by extreme irregularity and uncertainty.

'He (age 30) describes himself as a three-branch man. His main business is that of jobbing plumber. The usual wages of a London plumber are said to be 9d. an hour, and the weekly hours of labour 56½ generally, in the suburbs, 53 in the "City", and large suburban firms. This plumber trusts to his local connexion, and the information supplied by comrades, for his jobs.

'When out of work he applies to firms, and sometimes to likely householders. Other resources failing he tries to earn a trifle as a porter at auction rooms, or wherever he can get a job for the time. He is not deft at paper-hanging, and it took him 14 hours to hang 9 pieces at 6d. a piece, with his own paste (costing 2¾d.). His tools, worth about 5s., would cost 30s. to replace. He is often unable to do a job because his tools have been pawned. There is nothing else upon which he can raise a loan. The interest charged is ½d. in the 1s. for each month, and ½d. for the pawn-ticket.

'The man is of strong constitution. His only illness since marriage arose from lead poisoning, due to the inhalation of ingredients of colour on a day when he resumed work with an empty stomach after two weeks enforced idleness. This was the occasion of his transfer to the hospital. The demand for beds led, as he asserts, to his premature discharge. Having no money to pay his fare, he walked home (3 miles), and the same night had two fits—his first and last attacks—attributed to weakness and fatigue.

'Meals. Breakfast 8 a.m. Tea, bread and margarine or fat bacon. Dinner 12.45. Bread and margarine. Two or three days a week, meat and vegetables or fish. On Sunday, when possible, suet pudding is added. Tea 5 p.m. Tea, bread and margarine. There is never supper. The man takes a flask of tea with him in the morning, and warms it where he is working; he carries his bread and butter. His dinner, bread and cheese, or bread and a rasher of bacon, at an eating-house costs 2d. or 4d. When he is working within easy distance he comes home to dinner.

'This house is situated near Loughboro' Junction in the S.E. of London, in a neighbourhood thickly peopled by the lower middle class, by artisans of small regular earnings, railway servants, etc. This family occupies the top or second floor. Its two rooms are well lighted and ventilated. The front room is the living room of the family and

the bedroom of the parents. The boys sleep in the back room, the girl
on a bed-chair in the large room. The rent is 4s. a week.

'The furniture and clothing are very scanty, but kept fairly (not
perfectly) clean. The best room has a rough carpet, a few cheap prints,
a chest of drawers, and a little American clock. Out of doors the man
wears an overcoat, which is warm and conceals a deficiency of other
clothing. Indoors, or at work, the overcoat is removed, and reveals
him in shirt sleeves. The bed-clothes are a thin counterpane, and little
more.'*

In such an atmosphere the unions failed to strengthen their organisation in bad times as the employers had done. Various attempts to
amalgamate and federate unions came to nothing. The employers,
realising that unemployment meant that the unions had more control
over their branches, took the initiative in setting up boards of conciliation. The unions co-operated; as the A.S.C.J. put it:

'As a means of securing privileges . . . or preserving those we enjoy,
strikes have lost their ancient power, no doubt largely due to the
formation of powerful employers' associations and federations
throughout the country . . . coupled with . . . the law throwing its
protective wing over the 'blackleg workman'.'

When the national conciliation board was started it dealt with very
many demarcation and other disputes, often successfully. The building
labourers were, however, kept out.

The labourers were separated by a wide gap from the craftsmen.
Their wages, when they had work, were only 70% of the latter's. For
this reason and because they were the first to be sacked in slumps, they
had always been difficult to organise in unions. Attempts had been
made earlier. The great lock-out of 1859 (see Chapter 4) had led to a
Builders' Labourers' Union with 4,000 members. But it was short lived,
as was twelve years later the General Amalgamated Labourers' Union.
This came into being as a result of another lock-out in which the
labourers won an increase of ½d. per hour to 5½d. It had 5,000 members
but could not stay the course. It was its secretary, Patrick Kenney, who
wrote 'On the 20th September, 1860, I left off buying beer and took
to buying books to improve my mind.' But he fell into the hands of
the Conservative Party, took to drink, and to organising blacklegs.

* Quoted in E. Royston Pike, Human Documents of the Age of the Forsytes,
Allen and Unwin.

However, during the trade union upsurge of the 'eighties and in 1889, the same year as the strike for the dockers' tanner, the labourers tried again with the United Builders' Labourers' Union. The Bricklayers' secretary said, 'You cannot do it. I myself tried to help Kenney. It is impossible to organise labourers.' All the same, the union prospered during the booming 'nineties, and reached in 1901 the hitherto unheard of figure of 12,000 members. It then affiliated to the new Labour Representation Committee of the T.U.C., before most of the craft unions did. Although it lost many members during the following depression, this labourers' union had come to stay.

Those years of depression were miserable ones for the building workers. There was one worker who left a vivid picture of them. He was Robert Tressall, a painter who wrote a novel called *The Ragged Trousered Philanthropists* which was first published in 1914 and is now available as a paperback. It is about the daily lives of a group of painters and decorators in Hastings around 1906. The main character, Owen, tries to explain socialism to them but they prefer to accept their lot in which they hand over the results of their labour to the employer and the rich. So he calls them philanthropists. The other characters, Crass the chargehand, 'Misery' Hunter the foreman, Rushton the director, Bert White the apprentice, Dawson the labourer, are unforgettable.

As the countries of Europe drifted towards the war of 1914-1918, labour and capital in Britain came into bitter conflict. The cost of living was rising and wages were not keeping pace with it. Hence in the years 1911-1914 there were big national strikes in most of the large industries, by engineering workers, miners, dockers, and railwaymen. In the last three the Government had to intervene. The building workers also played a part in this nation-wide conflict. They tried to take two steps forward but were unsuccessful.

The first of these was an attempt to get amalgamation of the unions into one big union. At this time there were no less than seventy-two unions, national and local, in the building industry and thirteen local federations. Experience of the bad time and knowledge of the unions' weakness led more and more workers to call for industrial unionism, one big union for each industry. A 'Provisional Committee for the Consolidation of the Building Industries Trades' Unions into one Industrial Organisation' started the campaign in February 1911. Its leaflets called for a new union, to be called 'The Building Workers' Industrial Union which, in conjunction with other industrial unions, will ultimately form the framework of the machinery to control and

regulate production in the interests of the entire community . . . ,
maintain a fighting organisation' and would carry out 'the systematic
organisation of propaganda among the workers, upon the necessity of
being organised upon the basis of class instead of craft'. The campaign
became a crusade by rank and file union members, largely against
their own leadership. The employers took note; at a meeting in
November:

'They formed the London & District Employers' Federation, with
the object of defending themselves against the Socialistic Labour
movement and the offensive strike operations. This . . . is a reply to
the recent trend of action among the workers who desire to bring
about a General Federation of the Trade Unions in order to be in a
better position to war with capital.'

The employers need not have felt alarmed. On a ballot there was a
very large majority for amalgamation in principle, but the majority in
a second ballot on a concrete scheme was so small and in such a low
poll that it was nowhere near the two-thirds of the membership
required by law. A great opportunity was lost through apathy and the
desire of each member to keep the benefits paid by his union as they
were.

No sooner was this settled than the discontent among the workers
broke out into the great London conflict of 1914, the year war was
declared. Applications for wage increases were being slowed down by
the conciliation boards. The propaganda of the industrial unionists had
resulted in action to strengthen the unions by excluding non-unionists.
There were several unofficial strikes, in which the bricklayers took the
lead, to keep non-union workers off organised jobs. No doubt the
defeated industrial unionists wanted an outlet for their frustration.

Two days before Christmas, 1913, the London Master Builders'
Association issued a demand that the unions should punish their
members who struck unofficially against non-union labour, and
acceptance by 5th January. On 7th January, the demand rejected, they
gave all workers until 24th January to sign the following document:

'To Messrs......................
I agree, if employed by you, to peacefully work with my fellow
employees (engaged either in your direct employment or in that of any
sub-contractor) whether they are members of a trade society or not,
and I agree that I will not quit your employment because any of your

employes is or is not a member of any trade society; and I also agree
that if I commit any breach of this agreement I shall be subject to a
fine of twenty shillings, and I agree that the amount of such fine may
be deducted from any wages which may be due to me.

Witness..................   Name.......................

                                    Address....................'

About 25,000 workers refused to sign and were locked out. The
employers, who had reckoned to give a short sharp lesson in mid-
winter, were surprised when after two months hardly anyone had
signed the 'document'. They therefore retreated and agreed with the
union executives in April proposals for a settlement; these were that
the 'document' would be withdrawn, but all matters in dispute would
be submitted to the conciliation boards. In other words, freedom to
strike against non-union labour would not be allowed. The workers
rejected the compromise by 23,481 votes to 2,021.

The lock-out continued. By this time the electricians had joined in.
Five hundred members of the Electrical Trades Union in London
stopped work against non-union labour. Its Fulham branch wrote:

'We shall have to work too, with the blackleg and the scat, willy-
nilly, if the masters win; so, boys, keep a stout heart, fight and beat
the masters, then we will be able to say, "Oh" yes. We want to return
to work, but before we do you have got to send every scat and blackleg
on the streets as we have been.'

The breach between the union leaders (except for the bricklayers)
widened in May. They conferred with the employers who then
presented a 'final proposal'. In this the employers made further con-
cessions, such as permission to inspect union cards on the job, but
insisted there must be no strikes against non-union labour. Again the
workers rejected this by 21,017 votes to 5,824. But the stonemasons
voted for acceptance. The London employers then threatened to
extend the lock-out to the whole country. In June, faced with this
danger the union executives put yet a third, very similar, proposal to
the workers only for it to be again rejected.

By this time there was great hardship among the strikers, amongst
the labourers particularly. Families were evicted for non-payment of
rent and applications for assistance to the Poor Law Guardians received
little sympathy if the father was 'one of those trade union agitators

who were striking against their masters'. The E.T.U. stood firm with
them; its Central London branch reported:

'We are still in the thick of the fight. After nine weeks struggling
for better conditions our boys are as firm and true as in the first few
days of the strike and why should they not be? Have not the builders
shown us the way to fight for principle? Those brave fellows, the
builders' labourers, after 18 weeks of practically no strike pay, still
determined to stand out, then, surely we, who are supposed to be more
intelligent individuals, are not to be the ones to give in.'

But the strikers received a crushing blow when the stonemasons
broke away and negotiated separately with the masters. By early July
the masons were back at work. Then the employers announced that
the lock-out would became nation wide on 15th August which would
put half a million men out of work. The union leaders, dismayed, were
prepared to put pressure on the strikers by threatening to withdraw
union support. On 4th August Britain declared war on Germany and
a week later the strike was hastily ended. The proposals which the
men had rejected were now accepted. After seven months' struggle the
workers were defeated. They poured into the army and many were
soon fighting another battle in the trenches.

One result was, however, that in disgust, the leading militants
among the strikers themselves broke away and formed a new union.
This was the Building Workers' Industrial Union, a fighting organisation, without any friendly benefits.

FURTHER READING:

S. B. Hamilton, *A note on the History of Reinforced Concrete in Building*, H.M.S.O.,
1956.
T. J. Connelly, *The Woodworkers 1860–1960*, London 1960
J. O. French, *Plumbers in Unity*, London 1965
W. S. Hilton, *Foes to Tyranny*, A.U.B.T.W. 1963
R. W. Postgate, *The Builders' History*, N.F.B.T.O. 1923

The Building Trade Workers (1914-1939)

The fortunes of the building workers varied very much during these twenty-five years. The changes were brought about at different times by economic conditions, by war, by the policies of the trade unions and of the employers and by technical changes in the industry. The different housing policies of different Governments were also important. There were periods of years in which these ups and downs were marked: the first world war, the post-war boom, the slump from 1921 and recovery, the General Strike of 1926 and its consequences, the depression and recovery of the 1930s.

The following approximate figures of membership of trade unions suggest these changes. Behind these figures lay the well-being or its opposite of many thousands:

Trade Union Membership in the Building Trades

Date	Wood-workers thou.	Brick-layers thou.	Masons thou.	Plumbers thou.	Painters thou.	Plasterers thou.	Labourers thou.
1914	79	29	11	13	16	8	8
1919	134	49	13	20	38	—	65
1921	127*	75†		25	61	—	85
1927	117	59		21	38	12	35
1933	101	50		24	30	10	22
1939	151	65		29	47	11	—

* A.S.W. formed 1921.   † A.U.B.T.W. formed 1920.

When war was declared on 4th August 1914, most of the building trade unions were strongly pro-war and anti-German. Some, such as the National Association of Operative Plasterers, regretted the slaughter but called for a triumphant victory for Britain. Its general secretary wrote:

'Men who should all be working for one ideal, the brotherhood of man, are killing each other. And this is an age of so-called civilisation

and Christianity. The irony of it! Thousands of the poor are being
half starved whilst hundreds of the capitalist class are making huge
profits.'

But he concluded 'A victory for the cause for which we are fighting,
i.e. Civilisation and Humanity, is so necessary that we are compelled
to continue to the end.' Others, such as the large Amalgamated Society
of Carpenters and Joiners while recognising that 'business as usual
would be an absolute mockery', declared that to preserve their hard
won rights and privileges was a 'sacred duty' to the members in the
forces.

The war affected the different unions in different ways, according
to whether their trades were involved in the war industries or not.
House building came to a stop and membership of the masons' and
plasterers' unions fell, while the plumbers' increased only very slightly.
But in the shipyards and dockyards there were, for example, many
carpenters and joiners, and their union (the A.S.C.J.) grew quickly.
It came into conflict with the Government repeatedly in trying to
keep existing working conditions, all the more so as prices and profits
soon began to rise rapidly.

In 1915 the Munitions of War Act made arbitration compulsory in
disputes, suspended trade union restrictive practices and limited profits
in the munitions industries. It applied at first to engineering and shipbuilding, and the A.S.C.J. fought hard but unsuccessfully against
compulsory arbitration being extended to other industries, and particularly building. In the coalfields the miners defied the Act and in
July 1915 struck successfully in South Wales for an increase in wages.

In general there were few strikes throughout the war. The woodworkers, however, were involved in several disputes. They had found
a new branch of work in the aircraft industry. This became more
important as the Royal Flying Corps expanded into the Royal Air
Force in 1917. Production of aircraft was under the Ministry of Munitions with Lloyd George as its first Minister from 1915 to 1917, and the
second, Winston Churchill. The A.S.C.J. and the other woodworkers'
unions clashed with this Ministry over wages, piece-work, working
conditions and the employment of women. In September 1916, during the disastrous battle of the Somme in which British losses alone
were 420,000 men, the aircraft woodworkers struck for an increase
in wages. At the time of the battle of Arras, in the spring of 1917, the
union rejected a request for more piece-work from the Ministry and
secured a national agreement which allowed existing piece-work to

continue but forbade any more. The employers broke the agreement and the Government would not make them keep it. In February 1918, a month before the beginning of the final German offensive, the unions threatened to strike within nine days unless the Government forced the employers to toe the line. Their leaders were threatened with prison, but without effect. The Government therefore gave way and ordered the employers to apply the agreement.

Mainly because the union leaders had agreed to suspend the right to strike and because of compulsory arbitration the shop steward movement came to the fore throughout industry during the war. Action on the shop floor seemed to many workers the only way to get improvements or even keep pace with inflation. Before the war shop stewards had been collectors of dues; now they dealt with grievances and if necessary led unofficial disputes. In the building trades, shop stewards developed most among the woodworkers. Often engaged in war production as they were, the shop stewards needed more power to settle disputes at site or shop level. It was the stewards instead of the local branch which acted to protect working rules and conditions. By 1918 the A.S.C.J. found it necessary to make new rules for shop stewards which recognised these changes but laid it down that 'stewards shall co-operate with and be under the jurisdiction of, the management committee of the district'.

The war also brought about other changes in the trade union movement. On the one hand because the Government controlled the industry the unions often negotiated with Government departments. On the other, there arose the feeling that a better world should come out of the war and the Government appointed in 1916 a Ministry of Reconstruction to make plans for a better Britain. Although it issued many reports nothing much came of them. But these two conditions had effects on the unions generally and on the builders' unions in particular. Some of these were small, some large.

Thus it was in 1916 that under pressure from the Board of Trade the plumbers at last tried to settle the old conflicts with engineers about domestic hot water systems. They agreed a scheme of demarcation. 'It must be remembered', said the plumbers' council, 'that the Hot Water Engineers are a growing body, and the longer the delay, the more powerful our opposition.' But the agreement broke down.

More important was the attempt to improve the relations between capital and labour generally. In 1917 the Government set up joint industrial councils (Whitley Councils) which were to provide co-operation between employers and employed by increasing efficiency.

The main industries of Britain rejected the idea but it was tried in build-ing. The unions and the employers had already agreed among them-selves to a 'Builders' Parliament' which was to be the 'expression of a desire on the part of the organised employers and operatives to render their full share of service towards the creation of a new and better industrial order'. The 'Builders' Parliament' aroused great hopes. It was to deal with such matters as unemployment, employment of dis-abled soldiers, and technical training. The arguments for it stated at the time ran:

'The interests of employers and employed are in some aspects opposed; but they have a common interest in promoting the efficiency and status of the service in which they are engaged, and in advancing the well being of its personnel.'

In order to fit into the national scheme the Parliament's name was altered to 'Industrial Council for the Building Industry'. At its first meeting of 132 members there was much enthusiasm, but one delegate from the builders' labourers expressed doubts:

'Whilst one might have ideals it was necessary to recognise the things that one was up against. The building trade was not in itself master of the situation, but was co-related and intertwined with every other industry in the country. . . . While we had the present system it was useless to talk about artistry and idealism. But if one result of their meeting was to convert employers to the view that the present system was a rotten one and ought to be altered, then the Building Trades Parliament would do some good.'

In effect the 'Builders' Parliament' had no power and was doomed to failure. What it did do was to promote technical education and to give the workers a belief that they might obtain control of the industry. The slump after the war killed it and the employers withdrew in 1923.

A more lasting effect of the war was that the unions strengthened their organisation and took at last a step towards unity. The movement for amalgamation had been defeated before the war but still existed, as did the industrial union founded at that time, but was only just sur-viving. All the unions realised the need for greater unity but were not prepared to sacrifice anything for it. They thought they could get the best of both worlds by forming a federation. Thus in 1918 the National Federation of Building Trade Operatives (N.F.B.T.O.)

was set up and prepared to confront the employers. It soon started a campaign for the 44-hour week.

Two further steps towards unity also came out of the war and the Government's wish to deal with one union for each trade. Amalgamations were held up during the war by the absence of members in the forces but very soon after the woodworkers' unions united as did the 'trowel unions'. The Amalgamated Society of Woodworkers (A.S.W.)

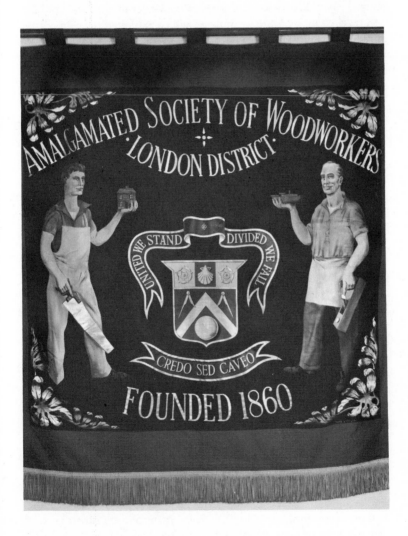

and the Amalgamated Union of Building Trade Workers (A.U.B.T.W.) were set up on the first day of 1921.

In the post-war boom during 1919 and 1920 the workers were well placed to press forward. There was an acute shortage of houses which the Government recognised by passing a Housing Act in 1919. This Act, for the first time, laid down a general subsidy for all council houses. There was also a grave shortage of building workers. As a result the new Federation had to resist the Government policy of dilution, backed up by a press campaign which exploited the cause of the ex-soldiers. It was able to insist that dilution should be gradual and only with proper training. On the other hand it pressed the Government to deal with price fixing and profiteering by merchants and builders which reduced the amount of house building; but it was not successful. The result was that the cost of houses rocketed and the scheme was cut short in the economy drive two years later.

The great achievement during the short boom in 1920 was the 44-hour week with the same pay as for the 50-hour week. This was the general rule although there were wide variations between 44 hours and 56 hours in different regions. The extraordinary thing was that it was given without a struggle. There were good reasons for this. It was not only that the employers were making big profits, but that the whole labour movement was on the offensive. The railwaymen's national strike defeated an attack on wages, the dockers under Ernest Bevin won a minimum wage, the miners and cotton workers struck, even the police formed their own shortlived trade union. To celebrate the 44-hour week the building workers took a holiday on 1st May, Labour Day. However, the danger of the shorter week was that many workers and employers saw it as an opportunity for more overtime. The N.F.B.T.O. therefore issued a ban on overtime after having appealed to the members:

'There are persons, who, having made huge profits during the war, are now extending their business premises and luxury buildings in preference to disgorging their gains by paying the excess profits duty. The few coppers extra for overtime would not affect these people: even were it doubled they would gladly pay it. It has a three-fold advantage for them. They get their premises up quick and start profit-making again. They dodge the Excess Profit Duty, and—from their point of view the greatest advantage—they succeed by insidious methods in securing the help of the workman to break down his own 44-hour week. It matters not how much they pay for extra time—they almost invariably get that back by increase of prices, but the increase of comfort and leisure secured to the operatives by a shorter working day is something they are unable to recover.'

For its part the carpenters' and joiner's union also cracked down on bonus systems. 'Even if we should lose a few thousand members now', they declared, 'we are determined that . . . it shall not be said that any member of our organisation is not a plain-time worker.'

The post-war boom came to an end in the winter of 1920 and a deep depression set in. By June 1921 there was a general movement by the employers for reductions in wages which had risen during the boom. One of the casualties of this turn about was the ship joiners who went on strike from December 1920 until August 1921 and then had to accept lower wages. Throughout the building industry wages fell until 1924, following the sliding scale agreed during the earlier boom. At that time, while the unions thought the sliding scale opened up a new peaceful future, some members feared that the unions might be weakened and the A.U.B.T.W. general secretary reassured them:

' "Whilst it may be argued that our organisations are kept alive by the uncertainty of wages", he wrote, "and by the adoption of a sliding scale the incentive is weakened, we want to state that in maintaining the 44-hour week, in resisting wholesale dilution, in fighting piece-work and payment by results, and in the complete stamping out of systematic overtime, etc., there is plenty of scope for the energies of our members." '

Very soon, however, the workers found they had to accept cuts. The wage of a fully qualified craftsman fell rapidly by 29% from 2s. 4d. an hour in 1920 to 1s. 8d. in 1922. In fact the standard rate did not reach the 1920 level for another twenty-six years.

The building workers made one big attempt to escape from the system of private enterprise which enclosed them. This was the National Building Guild formed in July 1921. Its beginnings went back to the war when labour wanted a real share in the control of industry. The urge for workers' control took various forms but one of them was a movement called guild socialism which made lively propaganda. Then in 1919 the 'Builders' Parliament' issued a scheme for reorganisation of the building industry in which the employer had only a fixed, not a speculative, profit and became in fact a manager on a salary in partnership with the operatives. This idea led the Manchester unions in 1920 to launch a Building Guild under their control and then to a network of guilds which united into the national one. Hopes were high. In the words of one bricklayer:

'By an organisation such as this we again become master craftsmen. We are the oldest organised craft in the country, and if these proposals can be carried through, we shall lay more bricks than private enterprise dreams of.'

The National Guild, which had help from the Co-operative Wholesale and Insurance Societies, undertook work to a value of over two million pounds. It was non-profit making, and when it built at low cost, as often happened, it handed back to the local authority the saving on the contract price. The workmanship and the houses were sound, carried out under good conditions of work and under the central control of the workers' representatives. The president of the federation of unions declared:

'Let me say for the Building Trade Operatives of this country that if the Municipalities will entrust their Housing to us and the Building Guilds, we will save them millions of money by a superior article and prove the fitness of labour for constructive government.'

But the National Building Guild had to close down in 1922. Like the Builders' Guild of Robert Owen ninety years earlier, it was a brave failure. The reasons were simple. Under the Housing Act after the war, builders could be paid for work as it was done and needed little capital. The Ministry of Health, responsible for housing, agreed to the Guild contracting for local authorities. Then came the slump, the Government axed the housing scheme and the Ministry withdrew its support. In the changed conditions the Guild needed a large capital if it was to survive. But as it had never made profits it had not accumulated any capital and could not get credit. It was wound up at the end of 1922, the movement faded away, and with it the smaller guilds in some other trades (clothing, furnishing, piano making) which had copied it.

Many workers had contributed their money to the guild and many went down with it. A member of the Plymouth Guild said:

'Do you realise that there are many working men who have given their last halfpenny into this movement? I, myself, have given every halfpenny. I have not a halfpenny in the world on purpose to pay wages, simply because finance is tight. . . . I have had no wages myself for six weeks. I am absolutely destitute. I had a matter of £200 saved up. Every halfpenny has gone into the Guild—everything—to keep the thing going. I realise the possibilities of this movement.'

The following years up to the General Strike of 1926 were marked by continual struggle by labour to regain some of the lost ground when trade improved and to resist reductions in bad times. The fact that trade was slightly better in 1924 and that the first Labour Government was in power, for nine months, encouraged the workers to press demands. In February of that year 100,000 dockers struck and won a wage increase, in March the busmen and tramcarmen struck, in April the shipyard workers, and in July the building workers.

They had been forced to accept wage cuts for two years. The employers had at one time actually demanded a reduction in wages of 20% together with an increase in hours from 44 to 47. When a lower cut was awarded by arbitration the general secretary of the A.U.B.T.W., George Hicks, thought it necessary to explain to members why the award had to be accepted:

' "As a product of negotiation the settlement is eminently sound", he reported to A.U.B.T.W. members. "The only argument that can be lodged against it is that of war at all costs, and no industry is at the present moment and within the limits of cold reason, in a position to wage war without counting the cost. The simple fact is that a lock-out or strike under existing circumstances could have no other result than compromise sooner or later, provoked by impoverished funds and the impetus of want and misery . . . compared with the economic position of other industries the building trades have not felt the full brunt of the capitalistic attack and are therefore stronger. This may be nothing much to boast about, but is a plain fact which indicates at least that the rout of the labour force has been partially arrested." '

The employers thereupon used his argument to increase hours of work. Therefore when trade improved in 1924 the unions immediately demanded an increase of twopence an hour. A settlement for one penny per hour was almost reached but the employers added further demands that the unions should give guarantees against breaking agreements. When these were angrily refused the employers threatened a national lock-out to 'enforce discipline and the observance of agreed working rules'.

The strike and lock-out lasted seven weeks. The A.S.W. paid out over £200,000 in strike benefits and the A.U.B.T.W. £11,000. It seemed as though the employers had wanted to stop building. Just at that time John Wheatley's Housing Bill was being piloted through Parliament, against some opposition. This Bill was the one successful measure passed by the Labour Government, which in any case did not have a majority in the House of Commons. It gave a subsidy of £9 a year to local authorities for every house built to let, and was the basis of council housing between the wars. Although Wheatley had carried the employers' organisations with him, the unions felt that the lock-out was meant as a blow against the Bill and they said it 'was admirably calculated to prejudice the Housing Bill, and our trouble has been used by the opponents of Wheatley's measure'. The Bill received the Royal Assent at the same time as the sittings of a Court of Inquiry into the dispute, which the Government set up because of the effect on house building. The final result was the same as the settlement nearly reached a few weeks before, plus summer working hours of 45½ instead of 44. The woodworkers were so angry with this result that their union withdrew for three years from the federation of building unions that had negotiated it.

Two years later came the General Strike in which the building workers took a full part. For them the issue was straightforward: the miners' standard of living was being attacked, and the mine owners were going back on national agreements. The mine owners, in their efforts to maintain profits, demanded a cut in wages and a repeal of the seven hour day. The miners replied: 'Not a penny off the pay, not a minute on the day'. The Trades Union Congress, realising that the standards of all workers were at stake, supported the miners, and after provocation by the Government called a general strike on 3rd May. A 'first line' of workers was called out, others being held in reserve. The strike was not in fact a general one; the 'second line', which included the engineering and shipbuilding workers, never came out. The first line included the building workers, along with railwaymen and transport workers, iron and steel workers and printers.

Some building unions, such as the A.U.B.T.W. and the A.S.W., were more militant than others and there was friction amongst them. This occurred over hospital and council house building. The T.U.C. had said that on these jobs certain kinds of work could continue. But the A.U.B.T.W. felt otherwise and declared:

'As we felt that the fight had to be short and sharp we were concerned more with finding excuses for bringing men out in support of the miners than in finding reasons for keeping them in. Under these circumstances our slogan "if in doubt, all out", was born.'

When the T.U.C. General Council called off the strike unconditionally on 12th May there was bewilderment, resentment and anger amongst the strikers. Members of the A.U.B.T.W. attacked its general secretary as 'one of the guilty men who had betrayed the workers by calling off the strike'. The woodworkers' union condemned the T.U.C. decision as 'an incredible surrender' and an 'inglorious and humiliating capitulation'. They called it 'one of the most deplorable and discreditable episodes in the history of trade unionism'.

For the miners were left in the lurch. They continued their struggle for six months until starvation forced them to accept lower wages and longer hours. The Amalgamated Society of Woodworkers made a levy on its members to help the miners and sent them £24,289, an average of about 3 shillings per member, approximately the rate for two hours' work. The Government, as eager for revenge as the mine owners, passed the Trades Disputes Bill of 1927. This made sympathetic strikes illegal and reduced the trade unions' power to spend money on political action by substituting 'contracting in' of their political levies for 'contracting out' as laid down in the Trade Unions Act of 1913.

After the General Strike both sides in industry wanted peace at all costs. They met to discuss production but nothing much resulted from the discussions. The economic situation and unemployment grew worse. The building industry however was not so badly affected as others; partly because of the house building by local authorities under the Wheatley Act. In this way the woodworkers, for instance, were able to carry the burden of their members in shipbuilding which became very depressed. But by 1929 unemployment in building rose sharply, because the Conservative Government cut the money for subsidies on housing.

In the general election of that year the unions, having lost the industrial struggle three years before, fought the political campaign to return a Labour Government to power. Their success was limited. Labour formed a government as it had the largest party in Parliament, but it lacked real power because it did not have a clear majority. There were many trade union members in Parliament (40% of the total), including no less than six from the Amalgamated Society of

Woodworkers, one of whom was a junior Minister. The building workers expected much from their members in Parliament. They were bitterly disappointed when the Government did not restore all the housing subsidies which had been cut by the Conservatives.

In fact the whole trade union movement was to be disappointed. In the general election the Labour Party had given a pledge to reduce unemployment but this the Labour Government failed to do. Unemployment rose and with it, the cost of benefits. When an international financial crisis occurred which threatened the gold standard of the pound the London bankers demanded reduction of unemployment benefit by 10%. The Labour Government split over this; a minority led by MacDonald joined with the Conservatives and Liberals to form a 'national' government. After this government had cut unemployment benefit and the pay of all State employees but failed to keep the pound on gold, in 1931 it held a general election and received an enormous majority. The overwhelming defeat of Labour in Parliament included all but one of the woodworkers' members. Their union now hoped for 'a big move forward for a militant working class party whose objective will be a rapid change of the present anarchic industrial system'.

But no such change came; unemployment grew. By the beginning of 1933 there were 2·9 million registered workers unemployed, nearly a quarter of the total; after that the number began slowly to fall. It was now the turn of the building industry to be the hardest hit. In 1932 the Government abolished the subsidies for council building, except for slum clearance. Council housing programmes were cut. At the worst time, out of about 103,000 bricklayers and masons over 40,000, 38% were out of work. A fifth of all members of the woodworkers' union were unemployed. They were prominent in the huge meetings of unemployed who demonstrated, to no avail.

The year 1933 was the worst for the building workers. Many of the tens of thousands unemployed were out of work for a long time. Union benefits had to be reduced. The State unemployment insurance benefits had been cut to 15s. 3d. a week for an adult man, 8s. for his wife and 2s. per child. These benefits could be drawn as of right only for twenty-six weeks. After that a transitional payment could be made but before men such as bricklayers and carpenters could receive it, they had to undergo the means test. This test was made on the means of the whole household, earnings of children, pension of grandparents, savings, were all taken into account and determined how much transitional payments was made. At one point a million workers were receiving these payments and enduring the humiliation of enquiries by the Poor Law authorities into their family circumstances.

When the situation improved in 1934 it was mainly due to the start of a new boom in house building. This time it was a private enterprise boom, the local councils had had their subsidies cut, and one which lasted until the war. Built mostly for sale to the middle class, great numbers of houses (over 1½ million by private builders in six years) were put up, creating new suburbs and a vast urban sprawl without any effective planning. They were not houses for building workers to live in. But by 1935 the men making them did have plenty of work; there was even a shortage of carpenters and joiners in some places and very few bricklayers were unemployed.

Full employment for the building craftsmen, but not for unskilled workers, came from the preparations for the second world war, though in Britain there were still well over a million workers unemployed when the war started. For several years before that Nazism and Fascism, as they spread through Europe, hung over Britain like a thunder cloud. In the fascist countries trade unions were suppressed in the interests of big business. The unions learnt through their international federations how workers in Germany and Austria were punished for organising underground trade unions. When the Spanish Civil War broke out in 1936 two thousand British went to fight for the Republic, five hundred of whom were killed. Among them were building workers, as well as miners, engineering and clothing workers.

Rearmament started, slowly, in 1936, and increased the demand for building workers who were already busy on house building. Later, with conscription, came the need for construction of camps and army hutments. Now, as before, during boom times, the unions faced demands from employers and the Government to drop certain trade practices and to accept dilution. But the Labour movement did not trust the Government's intentions during the period of appeasement of Germany. The building unions refused the demands; the woodworkers, for example, continued to oppose piecework.

Remembering their own history, the unions were determined not to weaken their standards and conditions of work before they were convinced it was absolutely necessary, as they became during the war. In the meantime they relied mainly on fair wages clauses in defence contracts, and their members to see that these were enforced. On one point, however, they had become less strict—entrance into the craftsmen's unions. For several years these unions, such as the A.S.W.

had been recruiting semi-skilled men who had not served an apprenticeship. They had done this in defence against the employers taking
on half-trained men for craftsmen's work. When therefore during the
immediate pre-war years there were complaints about labourers
being employed as carpenters for camp construction and accepted by
union branches, the union accepted the position.

This change in policy about recruitment probably came about partly
because of the technical changes in the industry. Although, generally
speaking, methods of production were much the same as they had
been in the nineteenth century there were certain changes in materials
and machines. The effects of technology on craftsmen's skills earlier
on have been described in Chapter 6. During the inter-war years
these effects increased. They were the result of manufacture of parts
and the use of steel, concrete and machines. The 'machine joinery
works' which had been complained of in 1905 (see page 166) had become
more numerous. Prefabricated joinery was frequently used. Also,
wood substitutes came into use, in the form of metal components and
plastics to a small extent. The Amalgamated Society of Woodworkers
faced up to these changes. Wood substitutes, it told members, which
required 'the use of joiners' tools in altering and fixing, should be
handled by carpenters and joiners. They can be so handled if our
members assert their rights *as soon as* the substitute is introduced'.
With more vision than before 1914 the union recognised the necessity
of changing itself with technical changes. In 1933 it gave a warning of
the threat from wood substitutes when it said:

'. . . it is not a pleasant outlook for our members, and these should be
faced with a resolute determination to adapt themselves to the use of
new materials, and *if necessary to modify the form of our organisation,*'

The rapidly increasing use of steel in construction had a different
effect, a new trade union. Quite early in the days of steel construction,
in 1913, a group of workers in Liverpool asked the then Steel Smelters'
Union to organise them. These men had worked on rigging and maintenance jobs inside steel works but now they were employed on the
erection of steel frame buildings as well as all sorts of bridge work.
Their skills were those of the steel erectors, riveters and rivet heaters.
The union was doubtful whether to admit them but did so and formed
several branches. It was able to raise their wages above those of labourers because of the skill and danger in their work. After the first
world war with the growth of steel construction, these men hived off

from the iron and steel trades to form the Constructional Engineering Union in 1924. Its membership was then 1,906. By the second world war, further expansion and new construction methods which increased certain occupations such as crane drivers, raised the number to nearly 10,000. Ten years before that the union had found its proper home by joining the building trades federation.

During the inter-war years there were other technical changes which affected all the workers in the building and construction industry. One of these was the coming into general use of concrete, the workers concerned being mainly organised in the Transport and General Workers' Union, along with thousands of building labourers. Another was the wider use on large sites of immensely productive machines such as bulldozers, trenchdiggers and concrete mixers. The men who operated these were classed as labourers and their wage was only a penny or two above the ordinary labourers' rate.

When the second world war started in 1939, it was no surprise, unlike the first war. The trade unions had condemned the Government's appeasement policy. When Czechoslovakia was about to be handed over to Hitler in 1938 they had declared: 'Whatever the risks involved, Great Britain must take its stand against aggression. There is no room for doubt or hesitation.' The building industry was fully involved in air raid precautions. The unions took part in the planning of rescue squads and demolition gangs, the construction of air raid shelters and aerodromes. They were now stronger than ever before and therefore better able to defend their standards of life and the homes of the people during the six years of war ahead.

FURTHER READING:

T. J. Connelly, *The Woodworkers 1860–1960*, A.S.W. 1960.
W. S. Hilton, *Foes to Tyranny*, A.U.B.T.W. 1963.
R. Postgate, *The Builders' History*, N.F.B.T.O. 1923.

8　The New Kind of Building

This chapter is about the work of a man who influenced deeply the new kind of building which first appeared at the beginning of the twentieth century and has become common since. It is also about the men who worked on the sites. The man was the German, Walter Gropius, who started the modern movement in building. The influence of Gropius can be seen today in almost any group of buildings or single large building. His designs meant a great increase in certain materials, particularly glass, concrete and steel.

WALTER GROPIUS (1883–1969)

Gropius was born on 18th May 1883. His father was an architect and senior civil servant in Berlin. His ancestors could be traced back to the seventeenth century in the neighbourhhod of Brunswick. They were middle class people, a variety of teachers, clergy, craftsmen, architects, painters and businessmen. Several of them were well known in the art and craft world. One, a manufacturer of tapestry, exhibited at the Crystal Palace in 1851. Another, who designed buildings and experimented with new materials, was principal of the Arts and Crafts School in Berlin and director of art education for the State of Prussia. Gropius's mother's family owned land in Pomerania and he spent his boyhood between the flat in Berlin and the wide, flat countryside of that region.

While Gropius was growing up the new German Empire, created 'through blood and iron' by Bismarck in 1870, was also growing in strength. Its industry and trade was increasing rapidly and becoming rivals to Britain. Between the years of Gropius's birth and 1900 German coal output increased three times, its steel four times, its pig iron three times, and, by 1914, had overtaken Britain. Even more remarkable was the growth of the chemical and electrical industries in which Germany led the world by the first world war. The electrical industry was dominated by two great combines, one of which, the A.E.G. (Allgemeine Elektrizitäts Gesellschaft) was to affect Gropius's training.

Thus almost overnight Germany changed into a highly industrialised country and the richest and most powerful on the continent.
The result was that the old handicraft industry shrank and the work of
craftsmen began to disappear. It was in order to try to find a new place
for design and craft in the industrialised society that there arose an
arts and crafts movement composed of designers, craftsmen, artists,
architects and builders. This movement began to get some support
from big business when Gropius was in his teens.

When he was twenty he went to the technical high school at Munich
for a year, and after a break of a year to the technical high school
at Berlin-Charlottenberg for two years where he studied architecture.
During these years he went on a long trip to Spain and worked in a
ceramics factory. While he was a student he completed the compulsory training in the army. It was also as a student, at the age of twenty-
three, that he designed in 1906 his first buildings. This first commission, was not surprisingly, in view of his mother's connections, for
several agricultural workers' cottages at Janikow in Pomerania.
They were severely plain structures designed with simple lines, without any ornament, and with small vertical windows.

The following year he entered the office of a Berlin designer and
architect, Peter Behrens. This man, who had started as a painter and
had been influenced by the English designer and socialist, William
Morris, had become well known for his modern designs of type faces,
table glass and buildings.

Two events occurred in 1907 which were important for Gropius.
An organisation called The Deutsche Werkbund (German Industrial
League) was founded, of which the aim was 'to raise the standard of
manufactured products by the joint efforts of art, industry and craftsmanship'. Gropius was a member and wrote for its yearbooks. In the
discussions which took place he began to work out his ideas of 'what
the essential nature of a building ought to be'. At that stage he praised,
for example, the grain silos of America as examples not merely of
containers of grain but as buildings whose shape and appearance were
models because they served their function perfectly and nothing more.

Also in that year, Behrens, Gropius's master, was appointed as consultant and designer to the big electrical combine, A.E.G., by its chairman, Emil Rathenau. Two years later, while Gropius was still in
his office, he built the turbine works for A.E.G., a large building which
was the most advanced of its kind in the world in its use of glass walls,
steel frame and clean simple lines. Behrens also designed many small
everyday articles such as street lamps and electric kettles on the same

principle which was also that of Louis Sullivan (see chapter 5) of 'form follows function'.

Gropius became Behrens's chief assistant. Just as Behrens was ahead of his time in factory and industrial design, so was Gropius in another field, housing. He saw that the only way to get good houses at prices and rents which working people could afford was through mass production. But it was standardised parts not complete houses which were to be mass produced. As early as 1909 he sent to Emil Rathenau a scheme for pre-fabricated houses. He wrote:

'. . . for economic production it was only by standardisation of component parts as distinct from the standardisation of complete houses that the architect could avoid monotony, provide variety and use factory production to the fullest extent.'

He explained how the house owner's desire for an individual home could be satisfied because of the infinite number of ways in which the interchangeable parts could be assembled. In this way also the designer or architect would be close to industrial production, as the Werkbund claimed he should be. Nothing came of the proposal at the time but Gropius persisted with it throughout his life.

After three years working for Behrens, Gropius, at the age of twenty-seven started in partnership with Adolph Meyer. They had been valuable and enjoyable years. The famous architect, Mies van de Rohe, who was also in Behrens's office at that time, described many years later a birthday party for Gropius: 'This party was in the back room of a very cheap restaurant in a suburb of Berlin. I remember this party very well. We had a very good time. I have never seen Gropius so happy. I think Gropius had the best time in his life. . . .'

A year later, in 1911, he designed a factory building which, in the words of van de Rohe, 'was so excellent that he became, at one stroke, one of the leading architects in Europe'. After searching all the periodicals for news of future buildings he had written to hundreds of industrialists suggesting that after his experience with Beherens he could carry out a project which would be 'both artistic and practical'. There was only one reply, from a manufacturer at Alfeld an der Leine, a town near Hannover, who wanted a shoe-last factory built. It was called the Fagus factory, from the Latin word, *fagus*, for the beechwood used in shoe lasts. Gropius's design was revolutionary in its time and today still seems contemporary. The exterior walls were almost entirely glass, with only thin brick piers between the expanses

of glass which was merely hung on the piers. The walls did not carry
or support the building and were only shields against the weather.
As Gropius said:

'The role of the walls becomes restricted to that of mere screens
stretched between the upright columns of the framework to keep out
rain, cold and noise.'

At one corner of the building the glass walls were joined directly
without any column, an unheard of feature at that time. Gropius justified this to the builder on the grounds of saving the cost of an unnecessary support. The owner was worried about the glare from sunlight
but loyally supported Gropius, and later on awnings were fixed in the
summer.

He followed this up in 1914 with buildings in which he developed
his ideas more clearly. At the League's exhibition in Cologne he showed
a factory building for a medium sized firm complete with office block
and car ports. In it he repeated his glass walls with continuous glazing
at the corners. The office block attracted most attention. The most
striking feature was that he brought two spiral stairways to the front
exterior at each corner and encased them wholly in glass. A contemporary wrote: 'A marvellous effect is achieved by two staircase towers
which he had enveloped entirely in glass. For window-gazing young-
sters these glass towers possess a number of attractions but other than
that they have no use.' In fact they led up to the roof where there were
covered dance floors and a restaurant.

Before war cut short his career Gropius also worked for industry.
For the German railways he designed a diesel rail car and a sleeping
car interior as well as furniture and wall fabrics. The 1914–1918 war
swept him at thirty-one years of age into the army as an officer in
the cavalry, and early on he was wounded by a bomb explosion. He
narrowly escaped death. Having invented a system of signalling he
was observer in an aeroplane when it was shot down and the pilot
killed. It was actually during the war that he received the invitation
which was to lead to fame. The Grand Duke of Saxe-Weimar, one
of the princes who still had great power and wealth in the German
states, asked him to become director of the School of Arts and Crafts
and of the College of Art in Weimar. This was on the recommendation of the famous Belgian designer who had established the school
of arts and crafts for the duke many years before. Gropius became
director in 1918 and in the following year he took the decisive step of

combining the two schools into a new one with a completely new
policy. This was called the Bauhaus, or house of building (its full
name was The State Bauhaus of Weimar). This new school became
famous and its influence spread throughout the world. Its aim was to
unite all the arts and crafts in the construction of fine buildings.

This aim was against tradition and it aroused much opposition
amongst conservative people. It was difficult to achieve in the
conditions of post-war Germany. Two facts illustrate the practical
and social difficulties. Gropius's contract of appointment as
director was made with the court chamberlain of the grand duke, but
with the consent of the provisional republican government of Saxe-
Weimar which had come to power as a result of the revolution against
the German Emperor after the war. It illustrates the continued conflict
which lasted for many years between the old pre-war Germany and the
men who meant to learn from the disaster of war and build a new life.
A bitter struggle broke out with local fighting repeatedly and led to
the Nazi Germany of Hitler.

The ideas of the Bauhaus were part of this conflict. Its first statement
ran:

'The complete building is the final aim of the visual arts. Their noblest
function was once the decoration of buildings. Today they exist in
isolation, from which they can be rescued only through the conscious
co-operation of all craftsmen.

'. . . Architects, sculptors, painters, we must all turn to the crafts.
Art is not a profession. There is no essential difference between the
artist and the craftsman. . . . Let us create a new guild of craftsmen,
without the class distinctions which raise an arrogant barrier between
craftsman and artist. Together let us conceive and create the new building of the future . . . which will rise one day toward heaven from the
hands of a million workers like the crystal symbol of a new faith.'

In 1919, a year after the opening of the Bauhaus, the following notice
appeared in Weimar:

'Men and Women of Weimar!
Our famous old Art School which we also want to develop, is in
danger.
'It is threatened because of the biased direction.
'Although certain people try to stop us speaking out about Art
we request all Weimar citizens to whom our art and places of culture

are sacred to attend a public demonstration on Thursday, 22nd January at 8.0 p.m.

'The elected committees.'

The 'biased direction' was a criticism of Gropius. Such criticism increased until after only five years the Bauhaus moved out of Weimar. Why did this happen? Weimar itself was a historic town with long traditions, where the great German poets had lived. But it was also there that after the revolution in 1918 a national convention met and drew up the constitution for the new German Republic. Thus the clash between old and new already existed, and the whole country was divided into right and left. The Government of the province of Thuringia which included the conservative town of Weimar, was socialist, whereas in Bavaria, just to the south, a communist uprising and soviet in 1919 was followed by a conservative regime. People in Weimar were already talking about 'art-bolshevism which must be wiped out'. Krupp and other industrial chiefs had already formed the Anti-Bolshevik League. Three years later Emil Rathenau, then foreign secretary, who was a Jew, was assassinated.

The Bauhaus soon made a reputation in spite of all the difficulties of inflation and poverty as well as violent political conflict. The students, numbering about 220, came from all parts of Germany and from Austria, and were of all ages from seventeen to forty. Most of them were men and many were ex-soldiers. As most of them also had to earn a living Gropius persuaded the authorities not to charge any fees. One student wrote:

'When I saw the first Bauhaus proclamation, ornamented with Feininger's woodcut, I made inquiries as to what the Bauhaus really was. I was told that "during the entrance examinations every applicant is locked up in a dark room. Thunder and lightning are let loose upon him to get him into a state of agitation. His being admitted depends on how well he describes his reactions." This report, although it exaggerated the actual facts, fired my enthusiasm. My economic future was far from assured, but I decided to join the Bauhaus at once. It was during the post-war years, and to this day I wonder what most Bauhaus members lived on. But the happiness and fullness of those years made us forget our poverty. Bauhaus members came from all social classes. They made a vivid appearance, some still in uniform, some barefoot or in sandals, some with the long beards of artists or ascetics. Some came from the youth movements.'

The teachers whom Gropius gradually brought together included some designers and artists who later became famous, such as Paul Klee and Wassily Kandinsky, but they were very advanced for that time and considered subversive by many of the good citizens of Weimar. In the preliminary year of the course, lasting half a year, each student was taught by a craftsman and by an artist. Following this, after being indentured, he went for three years into one of the workshops for instruction in a craft, either the carpentry, metal, pottery, stained glass, wall painting, weaving or sculpture workshop. The third stage was architecture including participation in buildings under construction. The syllabuses were based on instruction in materials and tools, book-keeping and estimating, and problems of form. These were dealt with first by study of nature and analysis of materials, secondly by geometry, construction techniques and drawing, thirdly by theories of design, space and colour. Throughout, the aim was not only to give a thorough training in skills but also to release the creative powers of the student. In this way the unity of the crafts and of art was to be brought about. In addition to the courses there developed very lively extra curricular activities, the students' dance band, dances in the local inns, revues and festivals involving flying of Bauhaus-made kites and lantern processions through the town.

Gropius had problems in arranging practical work for the architecture course. Land was available, but there was no money from the Government. Inflation raced ahead and the German mark had collapsed; people paid for their meals in billion mark notes. In 1923 one of the Bauhaus staff designed a one million mark note for the Thuringian Bank. On the land allocated by the Government, only one experimental house could be built for which Gropius raised money privately. He also employed the students on his own commissions which included a city theatre in Jena nearby and a large log house for a merchant at Dahlem, Berlin, complete with fittings and furniture. The land meant for a housing development was cultivated by the students to produce vegetables and fruit for the canteen. In order to keep this going Gropius sold a family heirloom, a silver table service and linen which had belonged to Napoleon.

Four years after the Bauhaus opened the Thuringian government asked it to organise an exhibition of its work for the public. By that time, 1923, political conflict in Germany had become more acute and inflation had reached its height. The French occupation of the Ruhr led to the right wing in Germany forming armed corps while in Bavaria Hitler assembled 5,000 men with arms taken from army barracks.

On the left wing, in Thuringia, the socialists brought communists into the Government. This was the setting for the Bauhaus Weimar Exhibition which fifteen thousand people visited during its three months. The exhibits aroused strong feelings, for and against, and the newspapers had headlines like: Storm over Weimar, Bauhaus Scandal, The Menace of Weimar, The Cultural Fight in Thuringia, Save the Bauhaus. The German League of Industry deplored the fact that 'the fury of political strife injected into all discussions of the work and purpose of the Bauhaus impedes any real consideration of the great and important experiment boldly going forward here'.

Soon the position became worse. The German Government in Berlin forced the left wing Thuringian provincial government to resign, while at the same time dealing gently with a right wing revolt in Bavaria. The enemies of the Bauhaus tried to prove that it was a left wing organisation, although Gropius had forbidden any politics in it. Houses were searched for evidence. Gropius recorded:

'Yesterday morning at half past ten I was called by a soldier from my office to my house because there was a search warrant. The house was searched by an officer and six men in a sensational manner. The order can only have been given on the strength of a malicious, irresponsible denunciation which was not checked. . . .'

Gropius was under great strain of both creating and defending his school. One of his teachers wrote: 'He works till three in the morning, hardly sleeps, and when he looks at me his eyes are like stars. I'm sorry for anyone who can't gather strength from him.' His business manager reported, 'Until recently it was possible to avert the most pressing dangers, but since the advent of the new government the official attitude, which had hitherto been indifferent, has changed into open animosity.' Even the local craftsmen became hostile; one newspaper cried, 'Bravo, Locksmith Arno Muller, for your telling words against the Bauhaus! How long . . . ?' At last in December 1924, the Bauhaus decided to leave:

'The Director and masters of the State Bauhaus of Weimar, compelled by the attitude of the Government of Thuringia, herewith announce their decision to close the institution created by them on their own initiative and according to their convictions, on the expiration date of their contracts, that is, April first, nineteen hundred and twenty-five.

'We accuse the Government of Thuringia of having permitted and approved the frustration of culturally important and always non-political efforts through the intrigues of hostile political parties. . . .'
(signed by all the masters)

All the students signed a similar statement.

Several cities wanted to have the Bauhaus but Dessau in the province of Anhalt, about 75 miles north-west of Weimar, was chosen on the initiative of its socialist mayor. Dessau was an historic town but it also had important modern industries—aircraft, chemicals, machinery, chocolate and sugar. The Bauhaus moved there in April 1925, with almost all the teachers and students, from whom five were appointed to the staff. It restated its first purpose: 'The intellectual, manual and technical training of men and women of creative talent for all kinds of creative work, especially building.' The city council approved a new complex of buildings which was begun in that autumn and finished in December 1926. Until then, the work carried on in a factory, a school and a museum.

The Dessau Bauhaus designed by Gropius was perhaps his most important building. It occupied an area of about 28,3000 square feet. One wing contained the advanced technical and professional department, classrooms, teachers' rooms, library and offices. From one end of that wing a bridge in which were the general administration and the architectural department, was carried on piers over the street. From the other end of the bridge on the right, another wing contained the workships and lecture halls; on the left, in a third wing, there were the big hall, dining hall and at its end a tall six storey block of students' study-bedrooms. The Bauhaus workshops themselves designed and carried out all the interior decoration and all the lighting. The materials and construction were as follows (from *Bauhaus*, edited by Bayer and Gropius):

'Reinforced concrete skeleton with "mushroom" columns, brick masonry, hollow tile floors. Steel window-sash with double weathering contacts. The flat roofs designed to be walked on are covered with asphalt tile, welded together, the tile laid on insulation boards of "tar-foleum" (compressed peat moss); regular roofs have the same type of insulation mentioned above, covered with lacquered burlap and a cement topping. Drainage by cast iron pipes inside the building. Exterior finish of cement stucco, painted with mineral paints.'

This gives little idea of the appearance of the buildings. The long horizontal stucco ribbons along the walls, the abundance of glass, the great glass curtain walls, the plain, simple openings, the rightness of proportions, made a startling impression of lightness, space and grace. There was nothing else like it in Europe. Although there was no attempt at symmetry in the old classical style, there was a curious reminder of Inigo Jones's Queen's House at Greenwich of the 17th century. There was the same use of cube shapes related to each other. In addition, Gropius designed and built in the grounds three double houses and one single one for the teaching staff.

The Bauhaus flourished with the city's help. A new department of typography and layout was added. Co-operation with industry increased; manufacturers accepted designs for furniture, light fittings, and fabrics and their technicians exchanged visits with the school designers. The Bauhaus got a considerable income in this way. The students became more active and had more of a say about policy. Originally there was a student council which Gropius consulted but at Dessau student representatives attended all meetings of the school council. It was also much easier to provide them with practical experience in their courses.

Gropius now received more commissions. First the City Employment Office, or labour exchange, in Dessau required for the increasing unemployed; a large half circle containing waiting and interview rooms, office block and extension shed for customers' bicycles. Then a workers' housing estate which in two years amounted to over 300 houses. He used standardised units but introduced variety by careful planning of terraces and five storey blocks. Next, came another housing project near Karlsruhe the 18th century town in the upper Rhine valley. The design won first prize in a national competition. Gropius co-ordinated eight architects' work on this scheme in which the buildings were a combination of four and five storey blocks of flats and terrace houses.

When the Bauhaus was well established at Dessau Gropius decided to leave. He explained why:

'I intend to leave the present scene of my activities, in order to exert my powers more freely in a sphere where they will not be cramped by official duties and considerations. The Bauhaus, which I founded nine years ago, is now firmly established. This is indicated by the growing recognition it receives and the steady increase in the number of its students. It is therefore my conviction (especially since my public duties are becoming more onerous) that the time has now come for me to turn over the direction of the Bauhaus to co-workers to whom I am united by close personal ties and common interests. . . .'

When the news spread during an informal party in the Bauhaus that evening the most senior student said to the director: 'You have made many mistakes, Gropius, but there is no one to fill your shoes. You ought not to leave us.' The students lifted Gropius up on their shoulders. It had been a long hard struggle for him and he wanted to do more building himself. He also felt his action would help the Bauhaus because the opposition had fastened on him. But since four outstanding teachers resigned at the same time there was probably more to it. One of them suggested this when he wrote:

'There must be room for teaching the basic ideas which keep human content alert and vital. . . . I can no longer keep up with the stronger and stronger tendency towards trade specialisation in the workshops. We are now in danger of becoming what we as revolutionaries opposed: a vocational training school which evaluates only the final achievement and overlooks the development of the whole man. For him there remains no time, no money, no space. . . .'

At the age of forty-five, he restarted his practice in Berlin. The only completed project during the next five years was a large estate of middle class flats at Siemensstadt, Berlin. The long clean lines of the four storey slabs and the fact that several different architects were each responsible for a number of slabs were typical of Gropius's work and his stress on teamwork. However, he designed many large projects which were not built for various reasons. Thus his grand designs for the Ukrainian State Theatre at Kharkov in the U.S.S.R., and for the Palace of Soviets in Moscow which was commissioned by the Soviet Embassy in Berlin in 1931, were never constructed because the Soviet Union had swung back against the modern style of men like Gropius.

Although many of his projects did not get beyond the design stage they had a great effect on later buildings. One of them, for a low cost housing project of three thousand dwellings at Spandau-Haselhorst, Berlin, was based on tall twelve storey blocks of flats. At that time (1931) these high slabs were far in the future, but Gropius did much to get people to see their possibilities and to accept them as a solution to the problems of land scarcity in towns by including them in different projects. Comparing them with houses and traditional flats he argued:

'The high-rise block on the other hand can be much more airy and sunny: there can be greater distances between buildings and large areas of parkland in which the children can have complete freedom to play and make as much noise as they like. . . .'

He was still working out these possibilities when he was in charge of a team of designers of the German exhibit at the Paris Exhibition of 1930. The exhibit was a community area inside a tall block of flats, and a single flat. The French who expected some heavy German work were deeply impressed by the lightness and grace. 'The first and most striking feature of the German exhibit is its lightness", wrote *Le Temps*. The German authorities had been nervous about the exhibit because it presented a new way of living in a new environment, but they were as pleased as the French.

However, in Germany, the economic depression became worse and the Nazis came to power. When they took over the government of Anhalt province in 1932 the Bauhaus had to move again from Dessau to Berlin. Hitler was appointed chancellor and head of the German Government in January 1933 and the Nazi terror began. All teachers, artists, designers, writers had to toe the line.

In April the Nazis closed down the Bauhaus which Goering called an 'incubator of cultural Bolshevism'. They used Gropius's Dessau building as a training school for their leaders. Other schools which had imitated the Bauhaus, such as those at Breslau, Halle, Stettin and Hamburg, also suffered.

The German Bauhaus was finished but it was not dead. The hundreds of men and women trained there were working all over Europe and its ideas and methods had influenced many thousands more. Later on in America they were practised in Chicago, New York, Harvard, California and South Carolina.

There was no future for Gropius and his colleagues in Nazi Germany. At the age of fifty-one he left to make a new career in foreign lands. The opportunity to leave came through his work on theatres. He had designed several theatres, the most notable being what he called the 'Total Theatre'. This was meant to draw the spectators into the play and get interplay between them and the actors. 'My "total theatre" ', he wrote, 'makes it possible for the actors to play during the same performance upon a deep stage, an apron stage, and within the central arena, or simultaneously upon all three.' In 1934 Gropius and his wife were given permission to attend a conference on theatres in Rome. From there they escaped to London in October.

Gropius spent three years in England. England could not give him much work but what there was had a big effect. The Bauhaus had made some impact and some modern designs had appeared. One of these was the Lawn Road flats in Hampstead where Gropius went to live. He entered into partnership with Maxwell Fry, of whom he wrote later: 'When I arrived in England in 1934 I found him at that early time in the front row of the valiant few who revolutionised architecture and building in England.' He had an interview at the Board of Education about the possibilities of employment but all they could suggest was a course of lectures at the Central School of Arts and Crafts. His command of English improved though at first his wife had to translate for him. But he was too advanced to be made director of the Cambridge School of Architecture as was then a possibility.

Gropius and Fry prepared many projects, one for Christ's College, Cambridge, another for flats at Manchester; but England was not ready for the modern style. Only two buildings were actually constructed. One was a private house of concrete and glass in Church Street, Chelsea, for a wealthy playwright. The other was far more important—the Impington Village College near Cambridge.

The Village College at Impington was one of eleven planned by the Cambridgeshire County Council in the nineteen-twenties, but only four were built because of the depression and cuts in expenditure on education. They were centres of cultural and social life in the rural districts for people of all ages. The Impington College, built 1936–1939, was a secondary school for 240 children between eleven and fifteen years of age, as well as a centre for adults. In it Gropius was free to express his philosophy.

The College was the first such building in England on modern lines and an example which was widely followed after the war. Until then schools had been built in a playground or round a quadrangle, or with an imposing front and a shabby back, often ugly and gloomy. In towns they were high buildings with small windows and dark corridors. The appearance of the Impington College was a complete contrast, though very familiar today. Long, low, single storey blocks joined at right angles, with large expanses of glass, open to the air and light, covered ways along the blocks, flat roofs, were the features which have become commonplace but gave a shock of surprise in the nineteen-thirties. All the shapes were strictly rectangular except the slightly curved block containing rooms for billiards and table tennis, a common room, a lecture room and a library. Linked to this were the

View of Impington Village College - Gropius (*Courtesy R. I. Severs Ltd.*)

refectory and kitchen, the assembly hall, with seating for 360, and stage, and the main entrance also curved and with a projecting canopy.

Before Impington College was finished Gropius had sailed for the United States where there were greater opportunities. But he had provided there the prototype for post-war colleges and schools in England. Not only was his open type of ground plan followed but also his ideas for industrialised building. Twenty-five years earlier, in 1910, he had proposed a 'Plan for Forming a Company to undertake the construction of Dwellings with Standardised Component Parts.'. Later, in the Bauhaus, unit construction was one of the problems studied. Then in 1927 Gropius built a house wholly made of factory produced components for an exhibition. Later still, after settling in America, he had developed by 1945 a complete system of prefabricated parts in the 'Packaged House System' of the General Panel Corporation. Every element was based on the same module of 40 inches.

In that same year in England, Hertfordshire County Council, faced with the need to build fifty primary schools in seven years, and a shortage of bricks, bricklayers and plasterers, decided on a big programme of prefabrication. This was very successful and the need was met which would have been impossible with traditional methods. The county architects were asked why a 40 inch grid was chosen. They replied:

'We adopted the standard recommended by Gropius, the British Standards Institution, the Ministry of Health and the Ministry of Education who advocate 40 inches on the ground of flexibility, with a long term aim of interchangeability of different manufacturers' components.'

Gropius had had his influence. And in another way as well. The Hertfordshire architects had to work very closely with the manufacturers and engineers so that they could understand the problems of industrial production. Gropius had argued for thirty years that this was necessary.

Very soon after his arrival in the U.S.A. Gropius was appointed professor of architecture in the Graduate School of Design at Harvard University, Massachusetts. He remained there as a teacher and leader of teachers for fifteen years. Although he continued his principle of unifying different subjects and disciplines so that specialists could understand each other and achieve teamwork he could carry this out only to a limited extent and slowly. Determined not to be cast as only a teacher he built a house for himself, with his partner from the Bauhaus, about 20 miles from Harvard. There, according to Siegfried Giedion, 'crowds of visitors used to come over every week-end, and often on weekdays as well, to see the newly finished "modern house"; for up to then not a single example could be found within a radius of upward of a hundred miles'. His first Government commission came in 1941 with a housing estate near Pittsburgh, for the workers in an aluminium factory. This was a defence contract, though before the U.S.A. was brought into the war by the Japanese attack on Pear Harbour. There were 250 units, mainly in thirty short terraces, of wood frame construction and end walls of buff brick. The terraces, carefully placed on a pleasant wooded site, made an attractive scheme but its bare simplicity roused a wave of criticism in the local press, which was a familiar experience for Gropius.

After the war Gropius formed a new team with seven young partners, The Architects Collaborative, and when he resigned from Harvard in 1952, he spent all his time with them. The T.A.C. opened up yet another career for the sixty-nine year old Gropius. It designed many large buildings and complexes: Harvard University Graduate Centre; U.S. Embassy, Athens; University of Baghdad, Iraq; Pan American Building, New York City; Tower East Complex, Cleveland; German Ambassador's Residence, Buenos Aires, as well as many houses.

When one of these, the fifty-nine storey Pan American Building above the Grand Central Station in New York, was under construction in 1960 Gropius was criticised for such a large building on a site in an overcrowded area, which ought to be a park. His reply was that that was a sentimental view since the area was already a vertical business centre and that he had in fact already persuaded the owner to accept 20% below the maximum square feet allowed by law.

Towards the end of his life the post-war Germany 're-imported' Gropius. As well as a glass factory he built Gropiusstadt (Gropius town), a 650 acre township in West Berlin for 44,000 people. In fact he had the difficult task of co-ordinating a group of Berlin architects who had been asked to keep to his principles but who had each followed his own ideas without paying attention to the overall plan.

When Gropius died at the age of eighty-six he had been active up to the last month. There was no funeral service, but at his wish his family and friends gathered at the Collaborative offices, drank champagne and read from his numerous writings. He is described as a man of medium height, wiry, energetic, with much humour and kindness, who hated pretension and status symbols, and had many friends.

As a teacher his aim was to give the student confidence, he guided with questions rather than gave instructions. He sympathised with the student unrest in the U.S.A. as he had with the student uprising in Berlin a few years earlier. When he spoke to the students there he did not preach at them but told them, 'it is always best to follow your own intuitions'. Talking to students at another time, he said:

'Act as if you were going to live for ever and cast your plans well ahead. By this I mean that you must feel responsible without time limitations, and the consideration whether you may or may not be around to see the results should never enter your thoughts. If your contribution has been vital, there will always be somebody to pick up where you left off, and that will be your claim to immortality.'

THE BUILDING WORKERS (1939–1960)

The basic question for the workers in the building industry during this period was whether they could maintain and improve their position in a world which was rapidly changing. These were years marked by the industrial upheaval brought about by a long war, the difficult changes required in post-war Britain and, throughout, technological advance which caused the use of new methods of work and materials. On the political side the Labour movement acquired new power and responsibility. This affected the building workers. Even the plumbers, the most non-political of all, decided at last in 1946 when there was a Labour Government to start their own political fund and affiliate to the Labour Party. This section is about how the building workers adapted themselves to meet the changing situation.

There was little change when war was declared in September 1939, until after May 1940 when the Churchill government was formed. The building workers were to play a great part. 'Britain', as W. S. Hilton says in his *Foes to Tyranny*, 'almost ceased to be a country. It became an aircraft carrier launching offensives against the enemy: a huge pillbox standing protectively against the nightly air raids— ready to resist invasion across the English Channel. And the transformation was achieved with bricks, mortar, concrete and steel. Aerodromes were laid down and coastal defences and air raid shelters built by men in the building industry. After the nightly destruction by

Hitler's bombers they were feverishly working round the clock, patching, repairing, ready for further onslaught.' But all that was ahead. The only effect at first was more unemployment. During the 'phoney war' months there was deep distrust between Government and unions. When the Government asked the T.U.C. how it would help the war effort it replied by requesting that the Trades Disputes Act of 1927 should be amended, and the answer by the prime minister, Neville Chamberlain, was that that depended on how the unions behaved during the war.

When, however, Churchill formed his government, the secretary of the Amalgamated Union of Building Workers wrote:

'The presence of Labour men in the Government, is evidence that the war will be kept to its liberating purpose . . . this is not the time to stand on ceremony or to tacitly follow the old traditions and formalisms . . . building workers, especially, need now to throw themselves, heart and soul, into the national effort.'

The chief 'Labour men' were Attlee, Greenwood and Ernest Bevin and, in a junior post, that same secretary.

There were problems, however, the two chief ones being dilution and piece-work. The fact that the Government took control of labour made these problems the more serious. The Government, through Ernest Bevin and by agreement with the unions, made arbitration compulsory and strikes and lock-outs illegal. Skilled workers had to register themselves and could be directed to 'essential work'. On the other hand the Government guaranteed that any craft practices which the unions gave up would be restored at the end of the war, as they had in 1914–1918.

The greatest need was for increased productivity. It was because of this, and also a sign of labour's increased strength that the unions made one great gain in the middle of the war. This was the annual paid holiday which started in 1943. They had also gained compensatory payment for broken time. In view of such advances the unions did not fight to resist piece-work. They were reluctant to lift the ban because they did not believe that payment by results would give increased production. However, there was no real choice because in 1941 the Government applied piece-work to all building work carried out under its Essential Work Order. The Woodworkers' Union had resisted most strongly and when it suspended its rule banning piece-work, soon after the German invasion of the Soviet Union, it went on record that 'the plain-time system is the only satisfactory method of payment for building trade operatives, and when hostilities cease (we) shall take steps to revert back to the system which is the fundamental basis of our organisation'. Soon, however, it was criticising the Government for not using fully its power over men, materials and money, for the war.

At the same time the unions gradually accepted dilution provided it was under union rates and conditions. The woodworkers agreed to unemployed plasterers doing certain carpenters' and joiners' work. Later on carpenters and joiners went into certain engineering jobs as dilutees. Women who were directed by the Government to woodworking processes were a special form of dilution. They were given only temporary membership of the union.

However, the crunch on dilution came towards the end of the war. The issue was important because the unions were asked to accept dilution for the coming years of peace. By 1943 the war was going well enough for people to think about post-war living. Britain had won Alamein and the Russians had destroyed a German army at Stalingrad; the Beveridge report and other plans came out. Among these was a plan for housing which would need half a million more men than there would be in the industry. How were the men to be found? The Government's answer was the national apprenticeship scheme plus another scheme for training 200,000 men for six months to enter the industry. The unions welcomed the first, there was much opposition to the second. As one union leader said: 'We can have serious unemployment for ten years or more and no one gives a damn for the men. But give us a few months in which labour seems unable to meet the demand and everybody wants to flood us with dilutees.' The unions did agree to the scheme but only on condition that the training of adults was controlled by the apprenticeship committees. Dilution for peace time had been accepted. To set against this, however, in 1945 when the end of the war was in sight, the workers gained a guaranteed week, of thirty-two hours, for the first time in their history.

The size of the post-war housing scheme, increased by the bomb damage, gave rise to arguments during the war about the proper standards of houses for the working class after it. The old view that bathrooms were a luxury still persisted, but it was scotched by a leading unionist in no uncertain terms:

'The time has passed when the workers could tolerate being stabled as a class apart, in any wretched shelter, at the behest of Governments, or lordly aristocrats, or any other members of the ruling class . . . and we call on the members of our Union, and the building workers generally, to do all in their power not merely to defend the best in existing housing standards but to improve them in every way.

When the Labour Government took over after the general election in July 1945, the woodworkers issued a warning:

'Paradise is not around the corner and there are many serious problems awaiting the new Government . . . wage problems, temporary unemployment in the war industries, housing, rationing, each bound to cause serious trouble.'

The warning was justified, but the Government and the industry set about the enormous task of making good the shortage of houses, officially estimated at 1¼ million in 1945. In fact 55,000 permanent buildings were completed in 1946, 140,000 in 1947, and well over 200,000 a year for the next three years, as well as 148,000 temporary prefabricated houses in that period, in spite of shortages of materials.

But in order to achieve that output various barriers had to be removed. One of these was the unions' hostility to incentive payments. As far as the woodworkers were concerned the problem first arose in shipbuilding where they had to agree to bonus schemes. But amongst the building workers there was widespread opposition, based partly on their feeling for craftsmanship. They were anxious to help the Government's housing programme but incentives were against their traditions. Their view was expressed by the president of the A.U.B.T.W.:

'When a man goes on to a job where the bonus mentality is supreme, he sometimes has to go through the degrading process of being sized up like a race-horse. Is he sound in wind and limb; will he have the stamina to last the pace of the bonus gang? If he hasn't these characteristics he can go to hell for all these people care. There are men in our industry who have, as trade unionists over a long number of years, given great service to their fellows and have assisted in improving their standards of life. Yet some of these men are amongst those who have been turned away from sites because they were not wanted by a bonus gang.'

There was a long debate. The unions held that the low output was due to the bad organisation in the industry. Nationalisation was put forward as a solution. The employers refused to give any wage increase without incentives. There was a press campaign against the bricklayers. Eventually in 1947, after a national ballot, incentive payments plus an increase of 3d. an hour were accepted; it was a break with tradition; in the old days men could be fined for undertaking piece-work. The effects, particularly of bonus schemes, were not so great as had been feared. One consequence was, however, the trend towards bargaining at the sites instead of nationally.

In the 1950s technical changes in the industry became faster. This was largely due to the war when there were many temporary buildings and when all building was based on the need to save labour and materials which were scarce. During those crucial years the most important effects were the use of light frames and claddings, pre-stressing of concrete, and standardised factory-made sections. This last led to the prefabricated dwellings. After the war certain materials increased considerably. The use of glass as cladding, of which Gropius's Impington Village College was a forerunner, of plasterboard and fibreboard are the chief examples. The new materials, new techniques and new methods of organisation were more to the fore in the construction of large buildings than in house building.

However, the unions were watchful of such changes and realised the impact they might have. The woodworkers were concerned about the increased mechanisation of their work and the prefabrication of building components. Many jobs had become so simplified that this could be done by process workers and had become 'irksome to the skilled operative'. Sometimes it was found necessary to claim them for the craftsmen at the stage of assembly or fixing. But the semi-skilled workers were part of the labour force and the question was whether they should be admitted to the union. The union decided that they should be, and women as well. At the same time the plumbers' union took the step of admitting plumbers' labourers, mates or assistants. It also changed because of technical advances like prefabrication of plumbing units and plastic materials.

New techniques and methods also led to demarcation disputes. There was sometimes a dispute as to whether the building agreement or the engineering agreement should apply to a site. The bricklayers in the A.U.B.T.W. were troubled with these problems and discussed at length what should be done. Various ideas were put forward, for example that apprenticeship and craft training might be altered so that there would be fewer crafts and the craftsman would be more flexible.

Some people thought that new techniques and materials were making the whole structure of the separate unions out of date. And in 1959 the National Federation of Building Trade Operatives in fact called for one big union for the industry. But this had happened before in the builders' history. The outcome depended on the views of each of the federated unions. The Amalgamated Society of Woodworkers, which was the largest, relying on its strength and the skill of its members, rejected the idea. The ideal of the one big union was buried again for the time being. The following year, however, the federation gained a major advance when it won a reduction in the working week to 42 hours with an increase in the rates to compensate for loss of earnings. And after this, when the idea of the one big union was brought up again in 1962, all the unions agreed to take steps which would make it easier to achieve unity of all the building workers.

FURTHER READING:

H. Boyer and W. and I. Gropius (ed.), *Bauhaus*, C. T. Branford Co. 1959.
J. M. Fitch, *Walter Gropius*, Mayflower 1960.
S. Giedion, *Walter Gropius, Work and Teamwork*, Architectural Press 1954.
T. J. Connelly, *The Woodworkers 1860–1960*, A.S.W. 1960.
J. O. French, *Plumbers in Unity*, 1965.
W. S. Hilton, *Foes to Tyranny*, A.U.B.T.W. 1963.
H. Pelling, *A History of British Trade Unionism*, Penguin Books 1967.

9   Social Control of Building

For at least eight hundred years men have found it necessary to control and regulate building. The reasons for this have varied over the centuries but the chief one has always been the danger to people's lives and properties from uncontrolled building. And this danger has in its turn arisen from the greed and anxiety to make money on the part of those landowners, developers and builders who acted without concern for their fellow men. Often the control has been exercised wisely, sometimes foolishly. Regulations have been made by royal decree, by Act of Parliament, and by by-laws issued by local authorities. Before Britain was an industrial nation, before the industrial revolution, they were simple and few, limited mainly to preventing fire. But in the nineteenth and twentieth centuries they grew more and more numerous and complicated as society became more complex.

We have already seen an example in this book, in Chapter 4, when the jerry builders in London were punished.

London was, naturally, where the regulations started. Even in the early middle ages, two and a half centuries after the Norman Conquest, it was a great city, by far the biggest in England. Before bricks came into use and where stone was scarce most houses were of wood frames. There was plenty of timber in the great forest just to the north of the city. Disastrous fires often broke out. 'The only plagues of London are the immoderate drinking of fools and the frequency of fires.' So in 1189 the mayor of London, Henry fitzAilwin issued a set of building regulations for the city but, unlike the later Building Acts, there was no means of enforcing them or of punishing offenders.

Party walls were to be no less than three feet thick and sixteen feet high. However, they could have alcoves one foot deep on each side, so that the real thickness could be only one foot. The minimum height of the overhanging upper storey of a house from the ground was eight feet. There was no rule about roofing materials then but about a hundred years later citizens were told to use lead, tiles or stone. When the chimney became more common during this period it reduced one kind of fire risk but brought another. It became necessary in the fourteenth century to order that chimneys should be made not of wood but of stone, tile or plaster. Sometimes a low stone wall was made in the hearth to protect timber walls where a large fire was required for baking and cooking and it was forbidden to place it near any partition wall of boards or lathes. The thatch on wooden rafters commonly used for roofing in London was one of the greatest risks. After a fire in 1212 in which London Bridge and the houses on it were burnt, building regulations prohibited the future use of thatched roofs. All the existing ones were to be plastered and any not so treated were to be pulled down. The mayor appointed inspectors—two carpenters and two masons, Thomas Mallynge and Richard atte Church. Bakehouses and brewhouses had special regulations; all unnecessary woodwork was to be removed and their walls plastered and whitewashed. In one street, Cheapside, all wooden houses were to be pulled down or altered to the requirements of the mayor and sheriffs.

Up to the Great Fire of 1666 London continued to be the main place for control of building. This was because under the Tudor and Stuart governments it grew so rapidly as a centre of industry, trade and government. There were three results. Disease and the death rate rose, the plague repeatedly took a heavy toll, notably in 1592, 1602 and 1603. Secondly, people felt that London devoured not only people but also the trade and industry of the ports and provinces of England.

In addition, the rulers feared its wealth and power as a rival to their authority, as the Civil War was to show. Therefore Elizabeth I, James I and Charles I tried repeatedly to limit the growth of the city but behind it were more powerful forces—overseas exploration, trade and conquest. In the same year that Drake brought home the *Golden Hind* with great wealth, Elizabeth issued a proclamation because of the 'plague of popular sickness':

'. . . all manner of Persons, of what qualitie soever they be, to desist and forebeare from any new building of any house· or tenement within three miles from any of the gates of the said Citie of London, to serve for Habitation or Lodging for any person where no former House hath been knowen to have been in the memorie of such as are now living; and also to forebeare from letting or setting or suffering any more families than are only to be placed or to inhabit from thenceforth in any one house that heretofore hath been inhabited.'

Three years later the Government, anxious about disease and law and order, had to remind the city that 'building had greatly increased

within the City and Liberties, to the danger of pestilence and riot',
telling them that offenders must be called before the Star Chamber and
that guilty workmen were to be put in prison.

As England's great conflict with Spain developed so did the struggle
about building between Queen and subjects. Four years after the
Spanish Armada an Act of Parliament stated:

'. . . no person or persons of what estate degree or condition soever
shall from henceforth make and erect any newe building or buildings,
house or houses, for habitation or dwelling, within eyther of the cities
of London and Westminster or within three miles of any of the gates
of the said City of London except it be to enlarge his or their house
or houses that so shall build the same; or to adde some other buildings
to his or their houses, or in their gardens for the more ease or pleasure
of the builder, or that such newe house or tenement shall be fit for the
habitation or dwelling of such a person as hertofore hath been assessed
to or for the subsidie to her Majesty at five pounds in goodes, or three
pounds in lands.'

The exceptions favoured the better off citizens; also sub-letting and
lodges were forbidden.

Enforcement was impossible and so in the last year of her reign the
Queen made another proclamation that in spite of the law:

'. . . yet it falleth out, partly by the covetous and insatiable dispositions
of some persons that without any respect of the Common good and
publicke profit of the Realme, doe only regard their own particular
lucre and gaine, and partly by the negligence and corruption of others
. . . that the said mischief and inconvenience doe daily increase and
multiplie . . .'

Therefore the Crown threatened:

'All tenements or buildings, not built upon the foundations of a
dwelling house, that are not at this present finished, to be plucked
downe and the builders or leasors to be bound to appeare in the
Starchamber.'

Such threats had little effect. In the next fifteen years there were at
least ten royal proclamations but there was no holding back the growth
of London. They made various commands. In one, buildings were to

'cause lesse waste of timber (fitter to be reserved for the shipping of this Realme)'. In another, for any new buildings 'the front and all the outer walls shall be built of brick or stone' and the front 'to be in that uniforme sort and order which shall be prescribed by the Alderman of the Ward or the Justices of the Peace'. In a third the beginnings of the building acts may be seen:

'Our expresse will and pleasure is, that in erecting of new Buildings hereafter the every whole story of and in such houses and Buildings, and all of every the houses of such whole story shall be of the height of tenne foot of assize at least . . . and if the such Buildings doe not exceed two stories in height, then the walles thereof shall be of the thicknes of one Bricke and halfe a Bricke's length from the ground unto the uppermost part of the said Walles: and where the Building shall be over the height of two stories, the walles of the first story shall be of the thicknes of two Brickes length and from thence to the uppermost part of the walle of the thickness of one Bricke and halfe a Bricke's length.'

It was obvious that the crown took this matter seriously when the king appointed a special high-powered commission to enforce his policy. It included the lord mayor, the attorney general, the chancellor of the exchequer, and Inigo Jones the royal surveyor. Jones with three others had power to punish offenders, which made him and the king much disliked by the London citizens. Another regulation specified the bricks to be used in London:

'That in the moulding of the said brickes the moulds be thoroughly and well filled . . . and that they be sufficiently and well dried before they be put into the kilne and thus carefully and thoroughly burned; so as for the assize, every bricke being burned, containe in length nine inches, in breadth foure inches one quarter and half a quarter of an inch, and in the thicknesse two inches and one quarter of an inch. That the price of brickes shall not excede the rate of eight shillings the thousand at the kilne.'

When the king had been removed the commission continued the same policy with its 'Act for the preventing of the multiplicity of Buildings in and about the suburbs of London and within ten miles thereof', and with Parliament's support it could impose a heavier penalty, of one year's rent.

In spite of these efforts, by the time of the Great Fire of London in 1666 the population of London had doubled in sixty years from 200,000 to 400,000. The fire burnt fiercely for four days at the beginning of September and destroyed four-fifths of the City. The government acted quickly, and appointed three commissioners, one of whom was Sir Christopher Wren, to survey the damage. By the following February the Act for the Rebuilding of the City of London had received the royal assent. This laid down complete building regulations. It specified the width of different classes of street and the type of the houses to be built in each. There were to be only four types of house (e.g. 'the first and least sort of house fronting by lanes') and the minimum thicknesses of walls in three of the types were also laid down. They were not regulated in the 'fourth and largest sort of mansion houses for citizens and other persons of extraordinary quality'.

The regulations were detailed, no timber was to be laid 'within the tunnel of any chimney'; the roofs, window frames and cellar floors and even the tile pins were to be of oak, instead of softwood imported from Norway. The one great advance on earlier regulations was that there were permanent surveyors to enforce them.

It was another hundred years or so before the next main London Building Act of 1774, a period in which the population of the city roughly doubled again. Such was the effect of winning an empire in India and Canada and of its trade and finance conducted from the City. So we find that one of the half dozen minor building Acts in that hundred years extended to cover for the first time St. Marylebone, Paddington, Chelsea and St. Pancras as well as the City and West-minster. In another Act we can see the impact of the great increase in traffic in and around London and the construction of turnpike roads to meet it; the Act for making the Euston Road turnpike required all frontages to be kept to a line of fifty feet from the footways. The legal size of bricks was also altered to 'not less than eight inches and a half long, not less than four inches broad, not less than two inches and a half thick'.

The main Act of 1774 was important because it controlled building in London for no less than seventy-one years, during the age of the industrial revolution and all the changes that brought about. Its chief objects were to prevent fire, to stop building encroaching on streets, and to give protection against dangerous structures and it went into the greatest detail about party walls, timbers, chimneys and permitted materials. It took a step forward by providing for district surveyors.

And it took a step back by abolishing the restrictions on the height of buildings which dated from the rebuilding after the Great Fire.

So far control and regulation existed hardly at all outside London. Important persons were at risk there more than elsewhere. As, however, the industrial revolution got under way, so towns throughout the land began to grow fast. About 300 towns appointed Improvement Commissioners, or Paving Commissioners, Lighting Commissioners, Police Commissioners, as they were sometimes called. Westminster set the example in 1765, Birmingham a few years later, and then town after town got the right to levy a house rate to pay for such improvements. But these benefited mainly the well to do citizens. In Manchester old town the commissioners paved the main and some of the by-streets, but the new town grew too fast for them:

'. . . single rows of houses or groups of streets stand here and there, like little villages on the naked, not even grass-grown, clay soil . . . the lanes are neither paved nor supplied with sewers but harbour numerous colonies of swine penned in small sties or yards, or wandering unrestrained through the neighbourhood.'

The standards of building could hardly have been lower:

'An immense number of small houses occupied by the poorer classes in the suburb of Manchester are of the most superficial character . . . . new cottages are erected with a rapidity that astonishes persons who are unacquainted with their flimsy structure . . . having neither cellar nor foundation. The walls are only half brick thick, or what the bricklayers call "brick noggin", and the whole of the materials are slight and unfit for the purpose. . . . They are built back to back; without ventilation or drainage and, like a honeycomb, every particle of space is occupied. Double rows of these houses form courts with, perhaps, a pump at one end and a privy at the other, common to the occupants of about twenty houses.'

The task was left to the speculative builder and the unscrupulous landlord, as at Bradford:

'. . . an individual who may have a couple of thousand pounds . . . wishes to lay it out so as to pay him the best percentage in money, he will purchase a plot of ground, an acre or half an acre; then what he thinks about, is to place as many houses on this acre of ground as he possibly can, without reference to drainage or anything, except what will pay him a good percentage for his money; that is the way in which the principal part of the suburbs of Bradford has sprung up.'

Throughout this time building was hampered by several taxes. Bricks, tiles, glass and windows were all taxed during this period. The duties on brick, which lasted for sixty-six years, till 1850, and on tiles for forty-seven years, till 1882, were light, though the penalties were heavy. But the tax on glass, which lasted ninety-nine years till 1845, was a heavy one. This discouraged the industry so that production of glass actually fell. More serious was the window tax which was not finally removed until 1851, the same year in which Paxton's Crystal Palace showed its blaze of glass. Towards the end of its life this tax was not payable on houses with less than eight windows and thus it was borne mainly by the rich. But the poor were also badly affected by it. In the northern towns many poor people lived in one large tenement house on which the tax had to be paid. A tax collector described the effect in Newcastle:

'No circumstance has contributed more to injure the habitations of the poor, and to diminish their healthiness, than the tax upon windows, the manner of its assessment, and the high duty upon window glass.... This heavy taxation naturally induced proprietors of such property to close up every window not absolutely necessary for light. Many of the staircases were so darkened that it became necessary to grope the way up them, at noon-day, as at night. The effect of this process upon ventilation was deplorable, and continues to operate to this day, for although the tax upon windows is considerably reduced, yet it falls heavily upon such houses.'

The result of such housing was disease and death for working people and the continual danger of infection for the rich. The plague, borne by black rats, was a thing of the past, but the cholera was not. This killing disease from Asia first made its way across Russia into Britain through Sunderland in 1831. It made a second visit fifteen years later killing fifty thousand, then again in 1853 and 1865 but with fewer deaths. It was because of this and also the efforts of a great reformer, Edwin Chadwick, that the struggle for public health succeeded.

This struggle started in 1838 when Chadwick, then secretary of the Poor Law Commission, a body much hated by working men, got three doctors to enquire into the causes of destitution and death in

London. The reason for the enquiry was the burden on the rates caused by the epidemics spread by insanitary conditions. As a result, two years later the House of Commons was moved to appoint its own Health of Towns Committee and two Bills were actually proposed, one to regulate buildings, the other on drainage. But nothing came of them. Then after another two years Chadwick published the famous Report on the Sanitary Conditions of the Labouring Population of Great Britain. The terrible facts it exposed stirred Parliament once again into life so that it set up a Royal Commission into the state of large towns and populous districts. Its report, mainly the work of Chadwick, came out in 1844 and 1845, and showed that out of the fifty large towns examined there were only six with good water supply and not one in which the drainage was adequate. But Parliament had other matters to deal with which it thought more pressing. In the meantime, certain towns did take note, for example, Liverpool, where the situation was, as described by a builder:

'From the absence of any systematic and compulsory arrangements, every man has built as it has pleased his own fancy, and little precaution has been taken as to drainage. There are thousands of houses and hundreds of courts in this town without a single drain of any description; and I never hail any thing with greater delight than I do a violent tempest, or a terrific thunderstorm, accompanied by heavy rain; for these are the only scavengers that thousands have to cleanse away impurities and the filth in which they live, or rather, exist.

'. . . the soil is subdivided into a multitude of holdings, and a man runs a new street, generally as narrow as he possibly can, through a field, not only to save the greater expense of soughing and paving, which, in the first instance, falls upon himself, but also that he may have a greater quantity of land to dispose of. The next owner continues that street, if it suits him, but he is not obliged to do so, and . . . the growth of narrow thoroughfares, the utter neglect of proper sewage, the inattention to ventilation, and that train of evils which is so much to be deplored, is the inevitable consequence.'

The town obtained an Act in 1846 which provided for 'effectually sewering and draining the borough', laying it down that no houses were to be created without drains and that every new street to be at least 30 feet wide, and also for the first medical officer of health.

The Government made a start with the public health act of 1848, but it was a weak start. Under the Act a town could if it liked have its own board of health to enforce proper drains and water supply, but the central Board of Health in London could only make it have one if its death rate was very high. After five years local boards existed for only two million of the people living in the towns. The countryside was hardly affected at all. The Board of Health was in fact abolished in a few years. This set-back was due to two things: public opinion, such as it was, disliked the central government having too much power over people and especially property, and secondly the personality of Edwin Chadwick who was secretary of the Board after leaving the Poor Law Commission.

Chadwick's drive had put drainage and sanitation on the map, but it then made many enemies for the new Board of Health. A very efficient man, he tried too hard to get the central government to push local councils towards public health. When it was suggested that he should give the town councils more scope he said: 'Sir, the Devil was expelled from heaven because he objected to centralisation, and all those who object to centralisation oppose it on devilish grounds!' He quarrelled with the doctors and with the engineers. They preferred large brick tunnels for drains but he insisted on narrow glazed pipes. When the Board was abolished he was retired with a pension of £1,000 a year which he enjoyed for thirty-six years.

Before then the first very small steps were taken towards decent housing. Parliament was persuaded in 1851 to make it compulsory for common lodging houses to be licensed and inspected. Another Act gave councils permission to build lodging houses but none of them did. Here is an example of the prosecutions under the 1851 Act:

'A house, No. 17, Lincoln Court, St. Giles, in one room 10 feet square, whereas three persons would be allowed by the regulations now enforced, seven men, nine women, and one child were found huddled together in a most filthy state; the bedding dirty beyond description, no partitions or ventilation; and a few minutes before the visit of the officer, one of the females had been confined. The keeper was summoned on the 24th October, 1854, to Bow Street Police Court, and fined £4 or six weeks' imprisonment.'

This was the year of the Great Exhibition. For that great display of British wealth the Prince Consort, president of the Society for Improving the Conditions of the Labouring Classes, had model houses built 'for the occupation of poor families of the class of manufacturing and mechanical operatives'. Each had a small lobby, living room of 150 square feet, scullery, parents' bedroom of 100 square feet, two child-ren's bedrooms of 50 square feet each, and a water closet. The construction as described in the Exhibition catalogue, ran:

'The peculiarities of the building in this respect are, the exclusive use of hollow bricks for the walls and partitions (excepting the founda-tions, which are of ordinary brickwork), and the entire absence of timber in the floors and roof, which are formed with flat arches of hollow brickwork, rising from 8 to 9 inches, set in cement, and tied in by wrought-iron rods connected with cast iron springers, which rest on the external walls, and bind the whole structure together; the building is thus rendered fire-proof, and much less liable to decay than those of ordinary construction. The roof arching, which is levelled with concrete, and covered with patent metallic lava, secures the upper rooms from the liability to changes of temperature to which apart-ments next to the roof are generally subject, and the transmission of sound, as well as the percolation of moisture, so common through ordinary floors, is effectually impeded by the hollow-brick arched floors. . . . The floors, where not of Portland cement, are laid with Staffordshire tiles, excepting to the right-hand room first floor, which is of lava. . . .'

The rents of 3s. 6d. to 4s. a week were to give a dividend of 7% on the capital invested. But there were few, if any, followers of the royal reformer whose model houses can still be seen in London at Kennington Park.

The return of the cholera fifteen years later again stirred Parliament into action; by an Act of 1866 local authorities could be at last com-pelled to remove nuisances and to provide sewers and water supplies. The change in public opinion which this required was largely the work of Dr. John Simon, chief medical officer of the Privy Council. His reports continually exposed unhealthy buildings and he had more influence than Chadwick had because he believed in working with the local councils. This is an extract from one of his reports:

'By places "unfit for human habitation" I mean places in which by common consent even moderately healthy life is impossible to human dwellers—places which therefore in themselves (independently of removable filth which may be about them), answer to the common conception of "nuisances"—such, for instance, as those underground and other dwellings which permanently are almost or entirely dark and unventilable; and dwellings which are in such constructional partnership with public privies, or other depositaries of filth, that their very sources of ventilation are essentially offensive and injurious; and dwellings which have such relations to local drainage that they are habitually soaked into by water or sewage; and so forth.'

Finally the Public Health Act of 1875 was important not only because it laid down a complete organisation throughout the country, including sanitary inspectors and medical officers of health, but also because for the first time sanitary authorities were allowed to make building by-laws if they liked. So far only London had been able to and had done this.

Throughout the whole of the previous hundred years there had been much suffering from bad buildings. But the worst victims were the climbing boys, the young chimney sweeps. The struggle to protect them from cruelty lasted for nearly a hundred years, and through no less than five Acts of Parliament between 1788 and 1875. The first of these mentioned the 'various complicated Miseries, to which Boys employed in climbing and cleansing Chimneys are liable, beyond any other Employment whatsoever in which Boys of Tender Years are engaged'. and said that no boy should be apprenticed before he was eight years old. Masters were to have apprentices 'thoroughly washed and cleansed from Soot and Dirt at least once a week' and 'in all things to treat his Apprentices with as much Humanity and Care as the Nature of the Employment of a Chimney Sweeper will admit of'. The Act was a dead letter; boys of four or five were still used and were choked to death. The Society for Superseding Climbing Boys agitated and got the invention of a satisfactory machine. But thirty years later the House of Lords rejected a bill to forbid climbing boys, in spite of evidence such as that given by a master sweep, who was asked if boys were sent up chimneys on fire. He replied:

'Yes, and rejoiced I have been many a Time, when such a Job has come to my Master, to get Sixpence for myself, or a Shilling; an active Child will not let the Fire rest on him; we pin the Bosom of the Shirt over, secure it in every Way, so that the Fire cannot get at him; we wet the Brush and then when one Boy is tired, we send up another, and if he keeps in Motion, the Fire will not lodge; if he is sluggish, he will be likely to be burnt. A Boy need never be burnt up a Chimney on Fire.'

The only practical difficulty in using machines instead of boys was the long horizontal flues which were put in the grand houses of the noble lords. These could not be swept by machines unless a door was made into the flue, but the objection was that such a door would spoil 'a handsome apartment'.

After another fifteen years, in 1834, there was 'An Act for the better Regulation of Chimney Sweepers and their Apprentices and for the safer Construction of Chimneys and Flues'. This laid down rules for the building of flues:

'Every chimney or flue not being a circular chimney of 12 inches diameter shall be in every section 14 inches by 9 inches; no angles less obtuse than 120 degrees.'

It also made it an offence to send a boy up a chimney on fire, but no one obeyed. So six years later another Act decreed master sweeps could not take apprentices under sixteen, but apprenticeship was going out, and there were plenty of unapprenticed boys to be had. Also the penalty for breaking the law was reduced from £100 to £10–£50. The boys' lot did not improve throughout the 1850s and 1860s in spite of the efforts of the Climbing Boys' Society and its chairman, Lord Shaftesbury.

Here is some evidence given by master sweeps in Nottingham to the Children's Employment Commission in 1863:

'At one time soon after the Act their number in this town was brought very low. But lately they have very much increased. There is a competition here between those who use boys and those who will not. . . . The law against climbing boys is a dead letter here.'

'The use of boys is much encouraged by the fact that many householders will have their chimneys swept by boys instead of by machinery. I have myself lost a great amount of custom which I should otherwise have. . . . I have been sent away from even magistrates' houses, and in some cases even by ladies who have professed to pity the boys, for refusing to use them.'

'No one knows the cruelty which a boy has to undergo in learning. The flesh must be hardened. This is done by rubbing it, chiefly on the elbows and knees, with the strongest brine, as that got from a porkshop, close by a hot fire. You must stand over them with a cane, or coax them by a promise of a halfpenny, etc., if they will stand a few more rubs. At first they will come back from their work with their arms and knees streaming with blood, and the knees looking as if the caps had been pulled off. Then they must be rubbed with brine again, and perhaps go off at once to another chimney. In some boys I have found that the skin does not harden for years. The best age for teaching boys is six. That is thought a nice trainable age. But I have known two at least of my neighbours' children begin at the age of five.'

'Nottingham is famous for climbing boys. This is on account of the chimneys being so narrow. A Nottingham boy is or was worth more to sell.'

Charles Kingsley wrote *The Water Babies* on the facts exposed by that Commission. The exposures made Parliament pass a Bill the next year by which master sweeps could be imprisoned instead of merely fined. Even two years after this there were still at least 2,000 climbing boys, mostly aged five to ten. And this went on into the 1870s. In 1875, after a sweep had received six months' hard labour for causing the death of a boy of fourteen, Parliament did make a stricter law with heavier penalties.

The control of building by by-laws throughout the country had been made possible in 1875, as mentioned on page 221. Before that time there was hardly any regulation outside London, which was a special case. London, of course, had its own building Acts. One of these, passed just before the cholera epidemic of 1846, was an 'Act for Regulating the Construction and Use of Buildings in the Metropolis and its Neighbourhood'. The growth of London was shown by the increase in the area covered—extending from Fulham in the west to Poplar in the east, from Wandsworth to Woolwich and from Hampstead to Greenwich. It said that the minimum width of new streets was to be 40 feet and that the height of buildings should be no more than the width of such streets. It made rules for drains and cesspits as well as for the ventilation and lighting of basements lived in. But the dice were loaded in favour of the landlords and builders. If a district surveyor brought an action against an offender which was unsuccessful he had to pay the legal costs out of his own pocket.

After another fifty years the London County Council took another step forward. It was high time. The Royal Commission on the Housing of the Working Classes had just shown that back to back houses still existed and the Census had revealed that 20 % of the people living in London were overcrowded, i.e. more than two to a room. Certain parts of London had been remodelled; for instance, the avenue named after Lord Shaftesbury had been driven across the slums of Seven Dials. But the London Building Act of 1894 did say that the open space behind any building must be in proportion to the height of the building, not merely in proportion to its frontage. This, however, applied only to new buildings. The evils of the old ones could be seen in the following written by Octavia Hill in her book *Homes of the London Poor*:

'The houses all round belonged to owners who had no interest in awarding a larger share of light and air to the dwellers in the court. Nor was there any means of compelling them to do so, since no Building Act lays down the amount of distance which must be allowed between the walls of buildings which have stood where they do now for many years. All that private effort unaided by statutory power could do to minimise the evil had been or might be done . . . but who among us could ever move back that great wall which overshadowed the little houses and made twilight at mid-day? Who would give space to move the water further from the dustbins, and the drains further from the ground floor windows? Who could remove the house at the entrance under which the archway passed, or that at the end, and let a free current of air sweep through the closed court? None of us.'

Outside London building laws had little effect for some years. Many places had no by-laws. But there was another way in which towns could improve their condition if they had the will to. This was the Artisans' and Labourers' Dwellings Improvement Act, also of 1875. Under it town councils could acquire insanitary areas by compulsory purchase, pull down houses, build new ones and in fact carry out an improvement as it was called. Few towns did this because it was too expensive on the rates, and even where it was done the real purpose of the Act, to provide housing for working men, was not achieved.

Birmingham was the outstanding example. Its improvement scheme was due to the drive of Joseph Chamberlain, future cabinet minister, who had become mayor in 1873. A crowded area of 93 acres was chosen, in which there were:

'. . . narrow streets, houses without back doors or windows, situated both in and out of courts; confined yards; courts open at one end only, and this one opening small and narrow; the impossibility in many instances, of providing sufficient privy accommodation; houses and shopping so dilapidated as to be in imminent danger of falling, and incapable of proper repair. . . .'

the fruits of which were:

'The evils ensuing were want of ventilation, want of light, want of proper and decent accommodation, resulting in dirty habits, low, health and debased morals on the part of the tenants.'

The corporation bought 43 acres at a cost of £1,310,000 and building started in 1878. But the centre and civic showpiece of the rebuilding was the imposing new street called Corporation Street with its highly commercial rents to set off against the cost. By 1882 when half a mile of the new street was open, 600 building including 375 houses had been demolished, 529 buildings had been thoroughly repaired and 2,263 dwellings temporarily repaired. But the question raised by a leading citizen was: 'May I ask whether any workmen's houses have been built, or whether the shops and houses created are not in the occupation of master tradesmen or other middle-class residents?' The answer was given in a local paper: 'It is little to the credit of the men who manage the municipal affairs of Birmingham that not one artisans' dwelling has been built out of the £1,800,000 which has been spent on the new street, and therefore that wretched and unwholesome dwellings, which still remain standing, are over-crowded to a fearful extent.' Families had in fact been evicted in the improvement area. The truth was that although the corporation had bought 14,250 square yards for working class houses, it had left the job to private builders, and none had been built. And none were until eight years later when the corporation did build 22 cottages and followed them up with another 81 houses. But this was a mere drop in the bucket of need. The economics of housing necessitated a council subsidy, but housing subsidies were then almost unheard of. The council refused any big housing programme because it would dis-courage private enterprise. This was the policy down to 1914. At that time Neville Chamberlain, the son of Joseph Chamberlain, declared, 'A large proportion of the poor in Birmingham are living under con-ditions of housing detrimental to both health and morals'.

Most towns believed that the solution to the housing problem was in the building of suburbs by private enterprise, and cheap transport to reach them. Because of this, the need for town planning arose. The suburbs of cities and towns did, in fact, grow fast. The expansion of

London's suburbs at the end of the 19th century has been described in Chapter 4. The same thing was happening, on a smaller scale, at all the large towns. It was made possible by workmen's trains, horse buses and trams—first horse trams, then steam, then electric. In the last twenty years of the nineteenth century the number of passengers carried by the suburban railways, bus companies and tramways serving London increased three times.

However, after a time there were misgivings about this sprawling, uncontrolled growth. In fifteen years half a million acres of agricultural land was used for building. The new suburbs were often oustide the control of town by-laws so that building, drainage and water supply were poor. Some of them were jerry built. Some, on the other hand, for the respectable middle classes, were built by co-partnership societies, as at Ealing (London), Liverpool and Manchester. When Hampstead Garden Suburb came to be launched the promoters stated, 'We aim at preserving natural beauty. Our object is so to lay out the ground that every tree may be kept, hedgerows duly considered, and the foreground of the distant view be preserved, if not as open fields, yet as a gardened district, the buildings kept in harmony with the surroundings.' It was to be a strictly controlled development which was necessary if the social aims were to be achieved. These aims were also stated in a publication called *Cottages with Gardens for Londoners*:

'We desire to do something to meet the housing problem by putting within reach of working people the opportunity of taking a cottage with a garden within a 2d. fare of Central London, and at a moderate rent. Our aim is that the new suburb may be laid out as a whole on an orderly plan.

'We desire to promote a better understanding between the members of the classes who form our nation. Our object therefore, is not merely to provide houses for the industrial classes. We propose that some of the beautiful sites round the Heath should be let to wealthy persons who can afford to pay a large sum for their land and to have extensive gardens.'

Needless to say, however, the 'industrial classes' did not move in.

Some years before Parliament recognised the need for planning or made laws to meet it, several pioneers planned and built model towns. Port Sunlight, Birkenhead, was one of the first. W. H. Lever, a grocer from Wigan, who became Viscount Leverhulme, had already made a fortune from 'Sunlight' soap when he started to build Port Sunlight

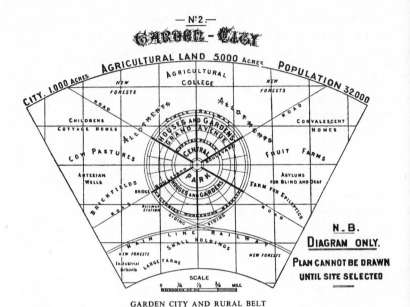

GARDEN CITY AND RURAL BELT

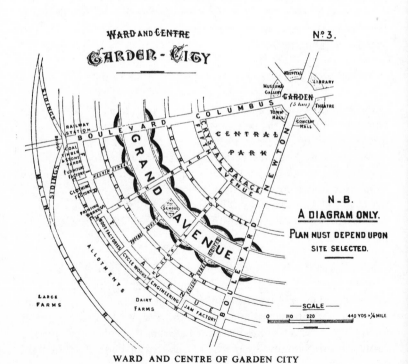

WARD AND CENTRE OF GARDEN CITY

in 1888. It was part of his management of his firm. When he opened the village two years later he said:

'Our idea, before we took the land at Port Sunlight, was that profit-sharing should be so managed that those who take the profits are those who are working at the works, and what we propose to do with the proportion of profit devoted to the workers is to apply it to the building of houses to be let at a reduced rental. We propose that those who have been longest in our service should have the greatest claim. We desire to encourage permanency in the people around us and we consider that by these arrangements the profits which are to be shared will always remain with the workers for the time being and not be divided up and squandered.'

The estate of 160 acres was planned for 1,500 houses and 10,000 residents.

As well as good, sound houses he provided schools, a church, social clubs, swimming bath, theatre, a technical institute. The whole place and its tenants were controlled by the firm.

Another pioneer, at almost the same time, was George Cadbury. After he had moved his factory out of Birmingham to Bournville for the sake of the country air he started his model village for the workers in 1895. But Bournville was far less under the control of its maker than was Port Sunlight. Houses were sold on long lease at cost price to workers who were helped to buy by loans at low rates of interest. From 120 acres in 1895 the estate grew to 672 acres twenty years later with a population of 4,390, and by the 1960s to 1,000 acres with 2,500 houses and 9,000 residents. There were plenty of open spaces, many more than at Port Sunlight, and every house had a garden. The roads had to be 42 feet wide and lined with trees and the houses set back a minimum distance from the pavement. The houses themselves were carefully and economically designed. A third businessman, Sir James Reckitt, a dye manufacturer, built Hull Garden Suburb. He stated his mixed motives quite openly: 'The only object in view is the betterment of our neighbours and to enable them to derive advantage from having fresh air, a better house and better surroundings. I urge people of wealth and influence to make proper use of their property, to avert possibly a disastrous uprising.'

After these businessmen came the great man of garden cities, Ebenezer Howard. While Bournville was under construction he published in 1898 his book, *Tomorrow; A Peaceful Path to Real Reform*,

later revised as *Garden Cities of To-morrow*. He was a Parliamentary
shorthand reporter who worked most of his life for the garden city
ideal and became, in recognition of his achievements, Sir Ebenezer.
His book began like this:

'The reader is asked to imagine an estate embracing an area of 6,000
acres, which is at present purely agricultural, and has been obtained
by purchase in the open market at a cost of £40 an acre, or £240,000.
The purchase money is supposed to have been raised on mortgage
debentures, bearing interest at an average rate not exceeding £4 per
cent. The estate is legally vested in the name of four gentlemen of
responsible position and of undoubted probity and honour, who hold
it in trust, first, as a security for the debenture-holder and, secondly,
in trust for the people of Garden City, the Town-country magnet,
which it is intended to build thereon. . . .

'The objects of this land purchase may be stated in various ways,
but it is sufficient here to say that some of the chief objects are these:
To find for our industrial population work at wages of *higher purchasing power*, and to secure healthier surroundings and more regular
employment. To enterprising manufacturers, co-operative societies,
architects, engineers, builders, and mechanicians of all kinds, as well as
to many engaged in various professions, it is intended to offer a means
of securing new and better employment for their capital and talents,
while to the agriculturists at present on the estate as well as to those
who may migrate thither, it is designed to open a new market for
their produce close to their doors. Its object is, in short, to raise the
standard of health and comfort of all true workers of whatever grade—
the means by which these objects are to be achieved being a healthy,
natural, and economic combination of town and country life. . . .'

More important than the book, however, was the way in which
Howard put his ideas into practice at Letchworth, Hertfordshire.
Largely by his enthusiasm enough money had been borrowed five
years later to secure nearly 4,000 acres at Letchworth for £155,587.
He and his supporters then formed a company, First Garden City Ltd.,
with an authorised capital of £300,000 to build the garden city. Its
prospectus stated:

'The object of the Garden City Company is, briefly, to erect a new
industrial and residential town of 30,000 inhabitants on an agricultural
estate, to avoid haphazard growth and the creation of slums by

building the town according to a predetermined plan; and to secure as far as possible the increased value of the land for the benefit of the inhabitants as a whole.'

The directors were mostly successful businessmen, including W. H. Lever and Edward Cadbury. They had a town plan made which allocated separate areas for factories, shops, civic centre and residences. Surrounding the town, and within the estate, was an agricultural belt. The company provided roads, water, drainage and public services, laid out the plots and leased them to public utility societies and private builders who built the shops and houses, mainly for sale. After the 1914–1918 war most of the houses were built by the council.

Letchworth did not grow as fast as was expected, but by 1914 the population was about 9,000 from a start of 400, and by 1945, 20,000. For many years the town was a much admired example of what could be done in planning a new town. The main interest for most people was the layout and the kind of housing, the tree-lined roads, the low density and the house gardens. At the beginning the housing problem was how to build cottages at rents which the lowest paid workers could pay at a time when there were no council subsidies and were not to be until 1919. The dwellings were therefore as cheap as possible. The rent of a cottage containing living room, scullery and three bedrooms was 5s. 6d. a week; a parlour added 1s. a week. This can be compared with the average weekly earnings of agricultural labourers in Hertfordshire of 17s. The rents were still too high for the lowest paid men, though Hitchin Rural District Council did manage to build a few cheaper cottages there. Some of the tenants' criticisms, according to C. B. Purdom, were on the following lines:

'Living-room. The largest rooms the most popular. More cupboards desired as a rule. Also shelves and kitchen dresser where absent. Objection to copper and bath in living-room.

'Scullery. Size of smaller sculleries found very inconvenient, especially with large families. Concrete flooring disliked, bricks or tiles preferred.

'Bath. Majority appreciate bath, but grudge space occupied in scullery. In two cottages visited with bath in lavatory tenants expressed great satisfaction with arrangement.

'W.C. When immediately inside back door some tenants complain of publicity and aerial communication with interior of house. Others

complain of publicity of outside door, but most tenants raised objection
to both positions.'

After the 1914–1918 war when the Housing Act of 1919 was passed
to provide 'homes for heroes' the houses built by the local council were
better in every way.

The open layout of roads and houses was a marked feature from the
start. The density worked out at 25 persons per residential acre. The
most characteristic sight, after a time, was the abundance of trees. No
less than forty-six varieties were planted in due course along most of
the roads, and these, with the greenswards, hedges and shrubs justified
the name of garden city.

The example of Letchworth helped to make people see the advantages of town planning. Older towns began to think that the planning
of suburbs was the answer to their problems. Birmingham City Council
came out in favour of it. And the Association of Municipal Corporations resolved:

'That power should be given to local authorities to prescribe and
regulate the planning of their areas in regard to the laying out of
streets in connection with building schemes or otherwise.'

At the same time a book called *The Example of Germany* (1904) made
known the legislation for town planning in many German states and
cities. This carried a lot of weight because Germany was the great
military and commercial rival of Britain and she had already set an
example in social welfare.

The result was the Housing, Town Planning etc. Act of 1909 which
made town planning schemes possible to secure 'proper sanitary
conditions, amenity and convenience'. The Government minister
responsible, John Burns, the former dock workers' leader, declared its
aims:

'Let us take Bournville for the poor and Bournemouth for the rich.
Let us take Chelsea for the classes and Tooting for the masses. What
do you find? You find in those four instances that your public-spirited
corporations and public-spirited landowners have been at work, and
I venture to say that if you take Bournville and Bournemouth, Chelsea
and Tooting, or towns like Eastbourne, you will find very much done
without damage to anybody of what we hope to make universal by
this Bill.'

Unfortunately that hope was not fulfilled. There were so many regulations in the Act that most local authorities did not try to apply it and after ten years, less than 10,000 acres had been dealt with. Thus when the boom in house building came in the years between the two world wars, there was no real planning to control it. During those years the fears expressed by John Burns in 1910 were fully realised:

'. . . may I bring before you—because it is my duty—the extent of the damage that is being inflicted upon rural England by the indiscriminate unorganized spreading, without control, of straggling suburbs?'

There were several planning Acts in those inter-war years but there was no real compulsion in them. It was not until after the 1939–1945 war that the first effective Act was passed in 1947 and in the previous year the New Towns Act.

In the meantime, however, Ebenezer Howard and his friends formed a New Towns Group in 1918. Out of this came Welwyn Garden City which was launched because it seemed hopeless to get a national policy on planning. Howard took the initiative in getting the land; he wrote:

'I knew that the reserve price was to be £30,000, and a few days before the sale I had only £1,000 in sight. That had been promised if I made a second effort to build a Garden City. I got busy on the telephone and by ten o'clock I had another £2,000 promised, two offers of £500 and one of a thousand. That was 10 per cent on the purchase price, enough to pay the deposit. So I went to the sale, made my bid and secured the site of the second Garden City.'

The land, plus two other purchases from the Marquis of Salisbury and Lord Desborough, came to a total of 2,378 acres at an average cost of £44. 10s. per acre.

Welwyn Garden City learnt much from the lessons of Letchworth, though these were mainly ignored in the country as a whole. There were differences between the two garden cities. The most important is shown by the 'Preliminary Announcement of a Garden City in Hertfordshire for London Industries', issued in 1919, which began:

A Satellite Town for London

The object of the company will be to build an entirely new and self-dependent industrial town, on a site twenty-one miles from

London, as an illustration of the right way to provide for the expansion
of the industries and population of a great city. Though not the first
enterprise of the kind (the main idea having already been exemplified
by Letchworth), the present project strikes a new note by addressing
itself to the problems of a particular city. To this end the site has been
carefully chosen so as to minimize the obstacles in the way of giving
a new turn to the development of Greater London.'

A new company, Welwyn Garden City Ltd., with capital of
£250,000 and directors from local notable and national figures, was
formed to build the town. It had difficulty in raising money because
the post-war boom was coming so an end. But it boldly stated it
objects:

'The town has been planned as a garden city with a permanent
agricultural and rural belt, and with provisions for the needs of a
population of 40,000 to 50,000. It will thus be seen that the scheme is
entirely distinct from a garden suburb, which by providing for the
housing of the people working in an adjoining district does nothing
to relieve congestion and transport difficulties.

'The maximum density of houses is planned for twelve to the acre,
and the average not more than five to the acre. The method of planning
proposed to be adopted by the company will not only tend to reduce
the cost of development, but will also preserve the amenities and
health of the town.

'In order to encourage the demand for sites and to stimulate the
rapid development of the town, the company is organized on the basis
of the original shareholders receiving dividends of no more than 7 per
cent per annum (cumulative). All further profits of the company
(subject to the payment of dividends on shares forming part of any
increase of capital) are to be expended for the benefit of the town or
its inhabitants. This expenditure will improve its amenities and tend
to lower rates and thus, it is believed, attract both residents and business
firms; and the better conditions so brought about, under which a large
working population will be living, cannot fail to promote their
contentment and happiness.'

The plan of the town then made divided the area into 1,298 residential acres, 170 industrial, 80 civic and commercial, 150 for schools,
608 for parks and rural belt, and 72 acres for the railway facilities. With

the industrial, civic and commercial areas at the heart the residential
areas were arranged around them. But some house building had to
come first so that industry might be attracted. The town had to be
developed section by section; when a road had been constructed the
houses were built on it. There were difficulties in getting houses
started, and so the company formed its own building subsidiary.
Houses to let at weekly rents could be built only by the local council.
The Government's housing policy was favourable to this during the
first years after the war but the council was slow to build houses; in
the first eight years only 443 houses. From then on the Government's
policy was to reduce building except for slum clearance and overcrowding so that house building at the new Garden City was handi-
capped. Up to the second world war the council completed six schemes
with another 699 dwellings. In one of them, as an example, there were
110 houses, 19 of which with parlour, 8 flats for old people and
12 garages. All the houses had three bedrooms and back-to-back fire.
'The construction was 11 inch cavity brick walls, faced with multi-red
facing bricks; the roofs were covered with non-tearable felt and red
pantiles; all roof spans were standard; the windows were good quality
metal casements, large and with plenty of opening lights giving well
lighted and ventilated rooms throughout. The doors and staircases
were in pine. The ceilings were of plasterboard and one coat of
plaster, walls two coats of plaster' (C. B. Purdom's *The Building of
Satellite Towns*).

Before long industry began to settle in the garden city, attracted by
the healthy living and working conditions, the labour available and
the rail and road connections. The Shredded Wheat factory was the
first important industrial building in 1924. The town grew at about
the same rate as Letchworth. The population, from a start of 430 in
1920, rose to 9,000 by 1931 and to 17,750 by 1946. The depression of
the early nineteen-thirties affected it, though not as much as many
places. All the same there was unemployment in Welwyn Garden
City and at one time 300 to 400 houses stood empty. The finances of
the company also suffered at that time. No dividends had been paid
on the shares because the cash available was ploughed back into
capital. When, in 1934 its finances had to be reorganised, a large
amount of capital was written off, and thereafter dividends were paid,
ranging from 2 per cent in 1936 to 6 per cent in 1947.

Welwyn Garden City was taken over by the Development Corporation in the following year but before then it had become the kind of
town, with its tree lined roads, that the promoters had planned thirty

years before, attractive to look at and healthy and pleasant to work and live in.

It was at the next and last big step in the social control of building that the Welwyn Garden City Company was taken over by the State under the New Towns Act of 1946. Responsibility for the town was thus transferred from shareholders to citizens. This Act was part of the new deal after the war. Many cities were shattered by bombing; planning would obviously be necessary. The authors of *A Plan for Plymouth* wrote:

'The immediate cause of the preparation of this comprehensive and positive plan for Plymouth is, without question, the destruction wrought by enemy action .. . . . Plymouth was no decayed or depressed area, no outworn town suffering from the aftermath of Victorian industrial prosperity and *laissez-faire*. . . . But like all old towns which have grown and prospered from small beginnings Plymouth was in need of a thorough overhaul—something more drastic than the preparation of a statutory planning scheme under existing planning powers which continually cramped the desires of the Corporation to bring their city up to date.'

There was no promise of 'homes for heroes' after this war as after 1914–1918, but as the authors of *County of London Plan* put it: 'There must be some plan of action to reward the valiant; works that can be put into immediate operation and will later fall into their ultimate place.' Rapid action was taken. The general election which returned a Labour Government took place in July 1945; the New Towns Bill was introduced in Parliament in the following April, and in the November the building of a new town at Stevenage, Hertfordshire, was announced. This was one of thirty-two new towns designated during the following years.

The New Towns Act said:

'If the Minister is satisfied, after consultation with any local authorities who appear to him to be concerned, that it is expedient in the national interest that any area of land should be developed as a new town by a corporation established under this Act, he may make an order designating that area as the site of the proposed new town.'

Each town had its development corporation, appointed and financed by the government, whose task was to plan and create the town. By 1970 fifteen new towns had been almost finished and a dozen more started.

The new towns were only one part of the post-war plans to improve the quality of living in Britain. The year after they were started the great Town and Country Planning Act was passed, great because it began a completely new system of development. It was also the first really effective Act. There were two chief reasons why this was so. County councils were compelled to plan their counties, and land-owners were compelled to sell land to them if required. Secondly, the Act dealt firmly with the old problem of development values, a problem which went back to the days of Lloyd George before the first world war. The question had always been, should not the nation rather than the landowner profit from the increased value of land caused by development? Now, the act gave to the nation all rights of development of undeveloped land; payment could be made to any owners who suffered hardship. Owners could not develop their land without the consent of the local planning authority. If, after getting approval, they did develop land, they were liable to pay a betterment charge.

Following this Act there were others which weakened it in some ways and strengthened it in others, but it was the turning-point. After it, building could no longer be carried out at the whim of the business man, builder, customer or local authority. It was now at least possible to plan and build new towns, remodel old ones, and preserve the countryside, for all the people. In fact more has been done in that direction during the last quarter of a century than in the whole history of this country, and in most difficult conditions. Fresh life has been given to an age-old ideal which had always been frustrated. Three hundred and fifty years ago Thomas More wrote down in his *Utopia* his vision of a town and country well planned:

'There be in the ilande 54 large and faire cities, or shiere townes, agreyng all together in one tonge, in lyke manners, institucions and lawes. Of these cities they that be nigheste together be 23 myles asunder. Againe there is none of them distante from the nexte above one dayes jorneyeye a fote. The precinctes and boundes of the shieres be so commodiously appoynted oute, and set fourthe for the cities, that none of them hath of anye syde lesse than 20 myles of grounde, and of some syde also much more, as of that part where the cities be of farther distance asunder. None of the cities desire to enlarge the boundes and limites of theire shieres. For they counte them selfes rather the good husbandes, than the owners of theire landes.

'The citie standeth upon the side of a lowe hill in fashyon almoust foure square . . . the stretes be appoynted and set forth very commodious and handsome, both for carriage, and also against the windes. The houses be of faire and gorgious building, and on the strete side they stand joyned together in a long row through the whole strete without any partition or separation. The stretes be twentie foote brode. On the backe side of the houses through the whole length of the strete, lye large gardens inclosed round aboute wyth the backe part of the stretes. Everye house hathe two doores, one into the strete, and a posterne doore in the back syde into the garden.

'They set great store by their gardens. In them they have vineyards, all maner of fruite, herbes and flowres. . . . Their studie and deligence herin commeth not onely of pleasure, but also of a certen strife and contention that is between strete and strete, concerning the trimming, husbanding and furnisshing of their gardens; every man for his own parte.

'And verelye you shall not lightelye finde in all the citie anye thinge, that is more commodious, eyther for the profite of the citizens, or for pleasure. And therfore it maye seme that the first founder of the citie mynded nothing so much, as these gardens.

'The houses be curiouslye buylded after a gorgious and gallante sorte, with three storyes one over another. The outsides of the walles be made either of harde flynte, or of plaster, or els of bricke, and the inner sydes be well strengthened with tymber work The roofes be plaine and flat, covered with a certen kinde of plaster that is of no coste. and yet so tempered that no fyre can hurt or perishe it, and with-standeth the violence of the wether better than any leade. They kepe the winde oute of their windowes with glasse, for it is ther much used, and somehere also with fine linnen cloth dipped in oyle or ambre, and that for two commodities.

For by thys meanes more lighte
commeth in, and the winde
is better kepte
oute.'

FURTHER READING:

N. Davey, *Building in Britain*, Evans 1964.
D. MacFadyen, *Sir Ebenezer Howard and The Town Planning Movement*, Man-chester University Press 1933.
C. B. Purdom, *The Building of Satellite Towns*, Dent 1949.
Schaffer, *The New Town Story*, McGibbon & Kee 1970.

Index